THE HOT TOPIC

WHAT WE CAN DO ABOUT GLOBAL WARMING

Gabrielle Walker and Sir David King

DOUGLAS & McINTYRE

VANCOUVER / TORONTO

Greystone Books
A division of Douglas & McIntyre Ltd.
2323 Quebec Street, Suite 201
Vancouver, British Columbia V5T 4S7
www.greystonebooks.com

David Suzuki Foundation
219–2211 West 4th Avenue
Vancouver, British Columbia · Canada V6K 4S2

First published in Great Britain in 2008 by Bloomsbury Publishing Plc.

Library and Archives Canada Cataloguing in Publication
Walker, Gabrielle
The hot topic: what we can do about global warming /
Gabrielle Walker and Sir David King.
Co-published by the David Suzuki Foundation.

ISBN 978-1-55365-371-4

1. Global warming. 2. Climatic changes. 3. Carbon dioxide mitigation.
I. King, D. A. (David Anthony), 1939– II. David Suzuki Foundation III. Title.
QC981.8.G56W35 2008A   363.738'74   C2008-900425-6

Front cover design by Jennifer Jackman
Cover photograph by © Dan Guravich/Corbis
Text design by Kaelin Chappell Broaddus
Diagrams by John Gilkes
Printed and bound in Canada by Friesens
Printed on acid-free paper that is forest friendly (100% post-consumer
recycled paper) and has been processed chlorine free.

For Rosa Malloy and Jane Lichtenstein

CONTENTS

PREFACE

The North Pole of planet Earth is an extraordinary place. It's a smudgy circle of frozen ocean, hemmed in by the surrounding landmasses of Siberia, North America, and Europe. Cracks occasionally appear in its surface where the ice has been torn apart by winds above and currents below. But for the most part, its gray-white facade is as unyielding as rock. You can walk on it, stamp on it, even land planes on it. When you're there, the Arctic sea ice doesn't seem remotely fragile, just motionless, silent, and strong, as if water had been turned irreparably to stone.

And yet photographs taken from satellites have now shown conclusively what scientists have been fearing for decades: The North Pole is melting. Each summer, the spread of the sea ice shrinks a little farther. It is vanishing from beneath the feet of the Arctic's polar bears. If we do nothing to stop it, by the end of the century the ice, polar bears and all, could be gone.

The story of global warming has progressed in the past few years from conjecture to suspicion to cold, hard fact. We now know for certain that on every inhabited continent on Earth, year by year and decade by decade, the world's temperature is rising. Something, or someone, is turning up the heat.

Should we care? After all, over the billions of years our planet has been around its climate has changed many times. In the geological past there have been ice ages, global floods, and heat waves. There have also been winners and losers throughout Earth's history—some species have become extinct while others have gone forth and multiplied.

But this time is different. If the current wave of change has its way with us, the polar bears will not be the only ones to suffer.

Human civilization has never before been faced with a climate that is changing this fast or this furiously. The threat has become urgent. In 2004 one of us (David King) caused a furor by describing climate change as "the most severe problem we are facing today, more serious even than the threat of terrorism."[1] Since then, the scale of the problem has become even clearer.

Also, the amount of material focusing on the problem has multiplied. Books, newspapers, TV, radio—another day, another headline. It has become almost impossible to sort out what really matters.

Amid this cacophony there is a handful of voices that persists in arguing that warming isn't happening, or that it's not caused by humans, while others see disaster around every corner and indulge in gory scenarios that have been labeled "climate porn." We don't agree with either of these approaches. Climate change is happening, and humans are largely to blame. However, we do not believe that disaster is inevitable. A few shiny new Priuses won't get humans out of this mess, nor will sticking our collective heads in the sand. But we still have time to tackle the worst aspects of climate change if we act fast and work hard.

In the course of this book we will pick our way through the blizzard of information and misinformation about global warming, explaining each point in the most straightforward way possible. We are both trained scientists, and our approach will be a scientific one—to examine the evidence, giving most weight to rigorous research that has been tested by peer review.

If you're looking for a debate about the science of global warming, you won't find it here, though we do cover some of the most common misconceptions about the problem in a handy list of climate myths at the back. What you will find is the latest scientific explanations of how much the world is warming, how we know that humans are to blame, what the worst-case scenarios might be, an overview of the most promising new technologies, and a political overview of where the world stands in its fight to solve the problem.

Though between us we have considerable experience in the worlds of media and politics, we are neither lobbyists nor politicians, and we have no personal axes to grind. We will lay out the entire essential story of global warming—what we humans have done, how we have done it, how we will need to prepare for the changes we can't stop, and how we can prevent the even worse effects that will otherwise follow. We aim to tell you everything you wanted to know about global warming but were too depressed to ask.

However, this is not a book about generic "green" issues. Most measures that increase efficiency and reduce waste will also help—at least a little—to reduce global warming. But this book is not a general environmental call to arms. It proposes a very specific set of solutions to a very specific, though wide-ranging, problem.

In particular, it seeks to show that the story need not have an unhappy ending. Global warming is a serious problem, probably the most serious that the human race has, collectively, ever faced. But we can still do something about it. This is a time for neither pessimism nor denial. It is a time for constructive, determined action.

PART I

THE PROBLEM

Before we can start discussing how to get ourselves out of the climate mess, we first need to set out the problem. There has been an extraordinary amount of confusion and misinformation about the science of climate change—which is surprising, since it is one of the few areas of complex science for which researchers are in almost unanimous agreement. In the next few chapters we will explain the science of global warming—what is happening, how we know the cause, the future changes that are now inevitable, and the ones that we still have a chance of avoiding.

1

WARMING WORLD

Climate change isn't new. Our planet is restless and its environment rarely stays still for long. There have been times in the distant past when carbon dioxide levels were much higher than they are today and Antarctica was a tropical paradise. There have been others when carbon dioxide levels were much lower and even the equator was encrusted with ice.

But over the past ten thousand years, the time during which human civilization has existed, Earth's climate has been unusually steady. We humans have become used to a world where the way things are is more or less the way they will be, at least when it comes to temperature. In other words, we have been lucky.

Now our steady reliable climate is changing, and this time nature isn't to blame. But how do we know for certain that the world is warming, and how can we identify the culprit?

The Heat Is On

When you're trying to determine whether the world's temperature is rising, the biggest problem is picking out a signal from the background "noise." Even in our relatively stable times, temperatures lurch up and down from one day to another, from season to season, from year to year and from place to place. To be sure that the underlying trend is changing, you need to take precise measurements from many different places around the world, and do so for an extremely long time.

We do have a few long temperature records, thanks to certain individuals who decided to make the measurements just in case they ever proved useful. The world's longest is the Central England Temperature Record, which is a tribute to the obsessive data-collecting habits of seventeenth-century British natural scientists. It covers a triangular region of England from London to Bristol to Lancashire and stretches back to 1659. This impressive record shows clear signs of warming, especially toward the end of the twentieth century.

However, the record covers only a tiny part of the globe. Changes in England don't necessarily reflect changes in the United States, say, or Brazil. It also doesn't go back far enough to reveal just how unusual our recent warm temperatures really are. How do they compare, for instance, to the apparent warm period in medieval times when the Vikings settled a verdant, pleasant "Greenland" and there were vineyards in northern England? Or to the so-called Little Ice Age in the midcenturies of the last millennium, when the River Thames in London froze over completely so that frost fairs were held on its solid surface?

To answer these questions, scientists have come up with ingenious ways to expand the records geographically and extend them backward in time. Some people have tried to interpret written archives that didn't quote actual temperatures,[1] but the best way is to look at records written not by humans, but by nature.

Every year, the average tree grows a ring of new wood around its trunk. In a good year the ring will be thicker, in a bad year, thinner.[2] Researchers drill a small core into the side of the tree, about the diameter of a wine cork, extract the wood, and then count and measure. By examining trees that are different ages, and even some trees that are long dead but have been preserved in peaty bogs, they have come up with a temperature record spanning more than a thousand years and from regions across northern Europe, Russia, and North America.

For more tropical regions, corals can play a similar role since they, like trees, grow a new ring every year. And in the frozen north and south (and the snowcapped peaks of tropical mountains), ice also contains a record book of past climate. Each year's snowfall buries the previous one. If temperatures are cold enough, the snow stays around long enough to be squeezed into ice, clearly marking out the annual layers because summer's snow crystals are larger than winter's, or because more dust blows in each year with the winter winds. The amount of snow that fell in a given year, and especially the changing nature of the oxygen atoms bound up in the ice,[3] gives clues as to how warm it was then.

Another clue comes from changing plant life, as written into the record of mud at the bottom of lakes. As temperature rises and falls, different plants flourish and each one sheds its pollen into passing currents of air. Some of this lands on the surface of a nearby lake, before slowly sinking into the mud beneath. Drill a hole in this mud, collect and analyze the pollen grains each layer contains, and you have yet another record of temperature changes over time.

Researchers have now used a host of different ways like these to analyze and splice together these different measures, and all come to strikingly similar conclusions for temperatures over the last thousand years.[4] The eleventh century was indeed relatively warm, corresponding to the Medieval Warm Period. ("Verdant" Greenland turned out to be more of a marketing exercise than the truth. Ice cores drilled into the heart of Greenland's ice cap show that a substantial quantity of ice has been present on the island for hundreds of thousands of years. Any Vikings who fell for the hype must have had an unpleasant shock when they arrived.)

Temperatures were also cooler in the seventeenth century, corresponding to the Little Ice Age, and again in the early nineteenth century. These warm and cool periods apparently were

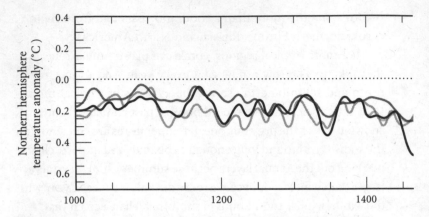

Changes in northern hemisphere temperature relative to the average value from 1961–90 in °C (1°C is approximately 1.8°F) for the past 1,000 years. The different lines reflect data that come from different sources and methods, but all show the same dramatic increase in temperature

also fairly widespread, though they may have been less preva-lent in the southern hemisphere.[5]

However, it was only in the twentieth century that tempera-tures really began to take off. The warming didn't happen regu-larly, but in two bursts—which turns out to be important. The first one occurred during the early years of the century and was marked enough that it made itself clearly felt. In 1939 *Time* mag-azine wrote: "Gaffers who claim that winters were harder when they were boys are quite right . . . Weather men have no doubt that the world at least for the time being is growing warmer."[6] But the following few decades brought slightly cooler tempera-tures, at least in the northern hemisphere, and public interest waned.

The second burst of warming began in the 1970s and has been gathering pace ever since. And, crucially, the temperatures we are experiencing now are hotter than they have been for the entire last millennium. Even the Medieval Warm Period was cooler than it is today.[7]

Let's look at some numbers. Globally averaged, from the 1910s to the 1940s temperatures rose by about 0.6°F. After that

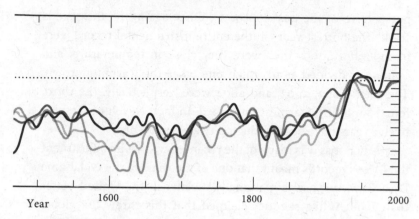

Year

1600   1800   2000

in the last few decades. (Source: P. D. Jones, T. J. Osborn, and K. R. Briffa, "The Evolution of Climate over the Last Millennium," *Science*, vol. 292 (5517), pp. 662–7, April 27, 2001)

there was a cooling of about 0.2°F, and since 1970 the world has warmed by a further 1°F.[8] These numbers might not sound like much, but they are very significant. Although the temperature where you live can change by much more than this within the space of a few hours or days, it is much more worrying when global annual averages show an inexorable upward trend. Averaging in this way smooths out short-term flurries and shows what's really happening. That's why a small change in global average temperature can reflect a very big shift in climate. Speaking in global averages, only a few degrees separate us from the frigid world of the last ice age.[9]

Though the proxy records of tree rings, ice cores, and the like give a good indication of average temperatures over a timescale of decades, they're not as accurate on temperatures for individual years. Thus, although we can say that the temperature is now greater than it has been in the past thousand years, it's harder to say how 2005 compared with, say, 1105. For that sort of pinpoint accuracy only a human record will do.

Good widespread records started to become available by about 1850, so we can put the past few individual years into the perspective of the past 160 or so. Once again, the message is stark. The hottest years in the entire instrumental record were in 1998 and 2005. They were very close in temperature, and opinion is divided as to which one takes the warming crown. The years 2002, 2003, and 2004 were, respectively, the third, fourth, and fifth warmest on record. In fact, eleven of the past twelve years have been in the top twelve on record.[10]

(Much fuss was made of the recent news that an adjustment to NASA's records meant that one of the years of the Oklahoma Dust Bowl, 1934, was marginally warmer in the United States than 1998. While skeptics claimed that this threw the global warming research into disarray, in fact it did no such thing. These two years were long known to be within a few hundredths of a degree of each other in the record of local American temperatures. But averaged over the whole world, 1998 and 2005 remain the joint record holders. Regional records can be interesting, but they don't tell the global story.)

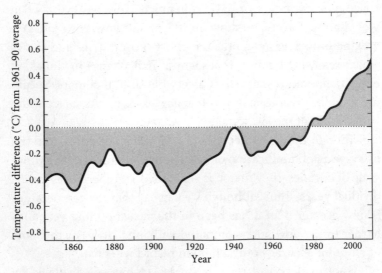

Temperature change over the past 150 years in °C compared to the 1961–90 average. (Source: IPCC)

The Intergovernmental Panel on Climate Change (IPCC) is an international body made up of leading climate scientists and government advisers from around the world. In 2007 it was awarded the Nobel Peace Prize for its work on identifying the causes of global warming. Because its reports must reflect the consensus view of all its many contributors, the IPCC has a reputation for being conservative. It is also widely considered to be the definitive authority on the science of climate change.

The latest report from the IPCC, published in 2007, describes the warming of the past few decades as "unequivocal."[11] There is no longer any room for doubt. The world is certainly heating up. What we need to know next is why.

People Who Live in Greenhouses

The prime suspect for causing this heating is, of course, the infamous "greenhouse effect." It is this that gives us our capacity to affect the climate; without it, we could burn fuel to our heart's content and the planet wouldn't feel a thing. Because of this, the greenhouse effect is often portrayed as the villain of climate change. But it may come as a surprise to learn that the effect itself is a very good thing. Without it, Earth would be completely frozen.

Looking at our nearest planetary neighbors, Mars and Venus, it's tempting to think that our planet has the best location in the solar system. Venus is closer than we are to the sun, and so hot that its surface would melt lead. Mars is farther from the sun, and its winters are so cold that steel on its surface would shatter. Earth, the in-between "Goldilocks" planet, is supposed to be just right.

However, that's not quite accurate. In fact, we're a little too far away from the sun for comfort. Going by our location alone, Earth should, by rights, be frozen over. This was discovered back in 1827 by French scientist Joseph Fourier, who had decided to try balancing Earth's energy books.

Our planet's central heating system ought to be fairly simple: Energy comes to us from the sun, in the form of sunlight. Earth soaks up this sunlight. It then glows with warmth, pouring out another form of light, which is too far beyond the red end of the rainbow for our eyes to see and hence is called "infrared." (All warm bodies give off this invisible infrared glow, including humans. Night-vision goggles work by detecting it, as do heat-seeking missiles.)

Fourier thought that the infrared light given off by invisibly glowing Earth would pour back out into space to balance the energy budget and set our global average temperature at 60°F. But when he calculated the heat energy coming in from the sun and going out from infrared radiation, he was astonished. By rights, our global average temperature should be a chilly 5°F. In other words, the entire planet should be frozen. Fourier also realized that each night, when incoming sunlight was temporarily switched off, the outgoing radiation would continue to pour into space, which should have cooled Earth's surface even more. In other words, there should also be much bigger differences between the temperatures of day and night than we see today.

Clearly something else is keeping us warmer than we deserve. Fourier realized that the atmosphere was the key. However, he didn't know which part of our air was acting as a warming blanket. This missing ingredient was discovered by a flamboyant Irishman named John Tyndall. Tyndall worked at London's Royal Institution, and when he wasn't wowing the audiences of poets and politicians upstairs with his famously entertaining lectures about science, he was down in the basement tinkering with his experiments on the atmosphere. He was fascinated by Fourier's calculations and wondered whether something was blocking part of the invisible infrared glow, preventing it from escaping back to space.

To find out, Tyndall set up an artificial sky in a tube and started shining infrared light through it. He wanted his sky to be as clean as possible, so he took out all "impurities" from the air. This left the two gases that make up more than 99 percent of our atmosphere: oxygen and nitrogen. But to his bafflement, infrared light slipped through the air unhindered. In other words, the gases that make up most of our atmosphere—nitrogen and oxygen—make no difference at all to its temperature.

On a slightly desperate hunch, Tyndall slipped a few of the "impurities" back into his air. He added a whiff of methane, some water vapor, and a soupçon of carbon dioxide, all of which exist in tiny amounts in the real atmosphere. And suddenly everything changed. As far as infrared was concerned, Tyndall's artificial sky went black. These so-called impurities did indeed trap infrared and prevent at least some of it from escaping back into space. They were Fourier's mysterious warming ingredients.

Tyndall and Fourier had discovered what we now call the greenhouse effect. Water vapor, carbon dioxide, methane, and the other so-called greenhouse gases share a special skill that the more abundant oxygen and nitrogen don't possess. They act a bit like the glass windows of a greenhouse, which allow sunlight through to warm the air inside but then prevent the hot air from escaping. The difference is that greenhouse gases don't block the air itself. Instead they're more like a one-way mirror. They let sunlight slip in through the atmosphere to heat the surface, but then block some of the outgoing glow of warmth that would otherwise carry heat back into space.

This discovery teaches two important lessons.

First, a little greenhouse effect is a very good thing. Without any intervention by greenhouse gases, our planet would be frozen and lifeless. Or, as Tyndall put it more poetically: "The warmth of our fields and gardens would pour itself unrequited into space, and the sun would rise upon an island held fast in the iron grip of frost."

Second, a little greenhouse gas goes a very long way. Watch out for people who say that greenhouse gases can't affect Earth's temperature because they make up such a tiny part of the atmosphere. They are certainly scarce compared to oxygen and nitrogen, but that's not relevant. Thanks to their skills at trapping infrared radiation, a whiff of greenhouse gas can change the temperature of the entire atmosphere, just as a few drops of ink can change the color of an entire bathful of water.

The real atmosphere contains a host of greenhouse gases,[12] but the most important are the ones Tyndall tried: water, methane, and carbon dioxide. Of these, gaseous water has by far the biggest effect on the temperature of the air. That's mainly because, compared to the other greenhouse gases, water is so abundant. It can make up anything from a fraction of a percent to several percent of the air, depending on the region, season, and time of day.[13]

However, when it comes to the power to change the climate, carbon dioxide and methane (and to some extent those few other, scarcer greenhouse gases) come into their own. Carbon dioxide makes up less than 0.04 percent of the air, and methane even less than that. But they both punch considerably above their weight when it comes to global warming, for two important reasons.

First, there is already so much water vapor in the atmosphere that human activities make hardly any difference to the total; rather, they are like adding a few bucketfuls of water to an ocean. But because there's relatively little carbon dioxide and methane, you don't have to add much to make a big proportional difference; adding a small amount is like putting a few extra bucketfuls of water into a bath. Thus, humans have already almost managed to double the amount of greenhouse gases in the air.

Second, by trapping extra heat themselves, these greenhouse gases also have an indirect effect on the amount of water vapor in the air. Warmer air can soak up more water, and warmer lakes, rivers, and seas can evaporate more easily into the atmosphere. The upshot of these two effects is that if you heat the air a little by adding extra carbon dioxide, it then takes up much more water vapor. This new water acts as a greenhouse gas in its own right and heats the air up even more, roughly doubling the effect the greenhouse gases would have had if they'd acted alone. Scientists call this a positive feedback—positive not because it's good, but because it amplifies the original effect rather than diminishes it.[14]

So, if you changed the quantities of the greenhouse gases in the air, you would expect the temperature to rise. And that's exactly what we have been doing for the past few hundred years. Since people began throwing coal on their fires, or stoking up their steam engines, carbon dioxide has been rising. The discoveries of oil and natural gas have accelerated the process. Every gas-fired cooker or central heating boiler, every tank of gasoline and every power station fueled by coal, gas, or oil has added a little more carbon dioxide to the air. Oil, coal, and natural gas are the three wicked witches of the climate change story, because they are all dense repositories of carbon.[15]

How Are Greenhouse Gases Changing?

Carbon is a wonder element, a supreme networker that can make chemical bonds with more or less anything else. Because of this extreme flexibility, carbon is the ultimate building block for all life on Earth. It forms the backbone for everything from carbohydrates, proteins, and fats to leaves, wood, bones, skin, and hair.

One consequence is that when you burn something that was once alive, you will release the carbon it contained, usually in the form of carbon dioxide.[16] That's what happens when you burn oil, coal, and natural gas. These three are known as "fossil fuels," because they are literally the fossilized remains of animals and plants that lived many millions of years ago and had, until the beginning of the Industrial Revolution around 1850, remained safely buried in the ground.

To make coal you need trees, swamps, and plenty of time. Most of the world's black coal was created between 360 and 290 million years ago in a period known, appropriately, as the Carboniferous. This was a time when insects grew to astonishing, horror-movie sizes—millipedes were six feet long, spiders and dragonflies spanned up to three feet, and even cockroaches could be as big as a foot. The coal itself came from the bodies of mighty trees growing in the swamps that covered the Carboniferous Earth.[17] When these trees toppled, the stagnant swamp protected them from rotting, and they were gradually compressed, dried out, and cooked into the coal that we burn today.

Making crude oil is a more delicate process. Oil was formed, much of it also in the Carboniferous period, from the dead bodies of tiny sea creatures, but these bodies had to be trapped, preserved, and cooked to exactly the right temperatures and pressures. That's why oil is found in far fewer places than coal. If the cooking process went awry, the oil broke down and turned into methane, also known as natural gas. Almost all oil deposits come with some methane, but methane is often also found on its own.

Since all three of these fossil fuels used to be alive, they are all made up of forms of carbon. Thus, burning fossil fuels produces carbon dioxide.

There's an important point to make here about the natural balance of carbon dioxide. Burning formerly living things isn't the only way that carbon dioxide can appear in the air. When we and most other living things breathe, we are "burning" our food to produce energy (which is one reason we talk about burning calories). And because our food used to be alive, the by-product is carbon dioxide. The carbon dioxide that we animals breathe out is taken up by plants, which use it to make their bodies, providing us with the food that completes the cycle. Vast quantities of CO_2 pass through the atmosphere perfectly naturally in this way every day. What's more, burning wood, crops, and anything else that used to be alive will also put carbon dioxide into the air. So why the focus on fossil fuels?

The reason fossil fuels are so important to the climate change story is that they involve burning carbon that had been buried for hundreds of millions of years. By contrast, the carbon inside wood was in the air fairly recently, probably just a few decades ago, before the tree soaked it up and converted it into trunk and branches. When you burn the wood, you're putting that same carbon dioxide back. In the long term, nothing changes.

Similarly, when you breathe, you release carbon dioxide that was probably taken up in the past year or two to make your food. Again, zero sum.[18]

But when you burn something that has been buried, and hence kept out of the air, for hundreds of millions of years, there's an important difference. By burning fossil fuels we are tapping into an old, deep reservoir that has long been locked away, and thus we are drastically changing the balance of the air.

Does that really matter? After all, as we hinted in the preface, there have been times in the past when carbon dioxide levels have been much, much higher than they are today. The trouble is that these times were unimaginably long ago, before humans were even a glint in nature's eye.

We know this because, thanks to a miraculous application of science and ingenuity, we actually have pieces of ancient air to study. These come from Antarctica, where the year-round freezing temperatures mean that snow that fell hundreds of thousands of years ago is still there today. It has been buried by subsequent layers, so you have to dig deep to find the really old snow (which has since been squashed down into ice). Researchers at Russia's Vostok station have done just that, drilling a core whose ice stretches back some 400,000 years. A subsequent core drilled at a joint French-Italian base at Dome C produced ice that is even older, at some 800,000 years.[19] Inside this ancient ice are tiny bubbles of air that were trapped when the snow first fell.

Scientists have analyzed these bubbles to check how much carbon dioxide and methane they contain. The results show that over long time periods, carbon dioxide levels naturally rise and fall. During ice ages, carbon dioxide tends to be quite low. During warmer times the carbon dioxide rises. However, there is not a single moment in the past 650,000 years when the amount of carbon dioxide has been anywhere near as high as it is today.[20]

We know that carbon dioxide levels have recently taken an ugly turn, thanks to an extremely persistent American researcher named Charles David Keeling. Back in the 1950s, Keeling decided to find out how carbon dioxide in the atmosphere was changing. He chose a site in Hawaii, far away from industrial sources that might bias his results, and then started measuring. In those early years he had constant trouble persuading funders that the study was worth doing, and, more important, worth continuing. But he succeeded, and his "Keeling curve" has become an icon of the global warming story.[21]

That's because the curve shows a never-ending rise. As the century has progressed, and we have burned ever more fossil fuels, Keeling's carbon dioxide curve has reared up like a malevolent serpent.

Once again, let's talk numbers. Because there's so little carbon dioxide in the atmosphere to start with, it's messy to use percentages. Instead, for carbon dioxide and other gases that are scarce in the air, scientists use "parts per million," or ppm. One ppm is 0.0001 percent.

During ice ages, carbon dioxide reaches a low of about 180 to 190 ppm. For the warmer periods in between (of which our current climate is one), carbon dioxide typically rises to a high of about 290. From the coldest point of the last ice age, about twenty thousand years ago, until 1900, the levels lay in a healthy range of 260 to 290 ppm.[22]

But over the time of the Keeling curve, the level has skyrocketed to about 383 ppm in 2007. The planetary atmosphere now contains a level of carbon dioxide that is nearly 40 percent higher than its "natural" preindustrial values, and it is still rising at between 2 and 3 ppm per year.

What's more, the levels of other greenhouse gases like methane have been rising, too. In the case of methane the reasons for the rise are more complicated—it comes from increasing the area of rice paddy fields, from gas that escapes when you're drilling for oil and even from belching cows.[23] In addition, there are the artificial chemicals, chlorofluorocarbons (CFCs), which nearly destroyed the ozone layer and happen to be excellent greenhouse gases in their own right. Putting all these together and calculating how much they add to the carbon dioxide effect, we now have a "carbon dioxide equivalent" (CO_2eq) level of about 430 ppm. In other words, we have effectively added about 60 percent to the greenhouse gases that were there before. And counting.

In summary, we know the world is warming. The physics first discovered by John Tyndall tells us that carbon dioxide is a greenhouse gas that heats Earth. The ice cores also show that when carbon dioxide goes up, temperature goes up. And carbon dioxide is higher now than it has been for hundreds of thousands of years. When you're looking for an explanation for the recent heating up of the planet, carbon dioxide and its sister greenhouse gases are clearly the likeliest culprits.

But did they really do it?

2

WHODUNNIT?

For the reasons already given in chapter 1, the prime suspect for the warming of our climate is, of course, the greenhouse effect. But there are still other, perfectly natural ways to change the climate. How do we know that this recent warming isn't just part of a normal cycle?

The best way to tell is to look at the characteristics of the heating, a fingerprint that identifies which mechanism is responsible. For instance, some warming mechanisms work geographically very widely, whereas others are more regionally focused. And some cause the atmosphere to heat through at every level, while others heat only closer to the ground. It's by looking at which possible mechanisms have changed in the "right" direction, and then using these sorts of arguments to choose between them, that scientists have reliably identified the culprit.

As we mentioned in chapter 1, Earth's temperature is set by the balance between the sunlight it receives and the infrared "heat glow" that it radiates back to space. So, in principle, there are four possible ways to turn up the heat:

Increase the Amount of Sunlight

Sunlight does indeed vary. Roughly every eleven years our parent star experiences bursts of energy followed by periods of sloth, and the strength and length of this cycle also change. For instance, during the so-called Maunder Minimum, between about 1645 and 1715, there were many fewer sunspots than usual, and this coincided with the coldest part of the Little Ice Age. Also, the sun was a little more active during the early parts of the twentieth century, and this probably contributed to that first burst of warming.

However, the sun has not been getting brighter in the past several decades. In fact, since 1970 the sun has actually experienced a slight cooling.[1]

Some researchers have suggested that the mechanism might be more subtle than this. Tiny, high-energy particles called cosmic rays constantly rain down on the earth from outer space. It's possible that when some of these particles hit the air, they encourage the formation of a cloud droplet. Thus fewer cosmic rays might equal fewer clouds, which—depending on the height of the clouds—might in turn mean that less sunlight is blocked.

The sun itself could conceivably help to bring this situation about. When the sun becomes more active, its magnetic field strengthens. This field stretches throughout the solar system and acts like a giant force field, blocking cosmic rays from entering our airspace. Thus, a more active sun would mean fewer cosmic rays, which might mean fewer clouds and a warmer climate.

This idea is intriguing, but it contains a lot of "mights," which troubles many researchers. A worse problem for its proponents is that recent evidence shows that numbers of cosmic rays haven't been falling lately, even though the temperature has certainly risen. In fact, like the sun's total output, cosmic rays have gone in exactly the wrong direction. If anything, they should have caused cooling.[2]

In other words, by whatever mechanism you choose, solar changes cannot explain the dramatic rise in temperature over the past few decades.

Reflect Less Sunlight Directly Back Again

The way to do this would be to change the planet's shininess. Some of the sunlight that arrives on Earth reflects straight back into space before it can do any warming. For instance, it can bounce back off the top side of clouds, which is why it feels cooler when a cloud passes between you and the sun.[3]

There's no obvious reason why clouds should have naturally changed in number during the last century. But sunlight can also bounce back if it encounters a haze of tiny particles, known collectively as aerosols because they float in the air.[4] If these have decreased in number in recent years, that could explain why the planet started to warm.

Natural versions of aerosols include bits of dirt, sand, or dust, volcanic ash, soot, or liquid droplets of sea salt, and even the largest particles are still smaller than the period at the end of this sentence. Most of these are whipped up briefly by the wind and last only a short time in the air before falling back to Earth. But the ash and droplets of sulfate put out by volcanoes are thrown up higher into the air, and can sometimes live long enough to cause significant cooling.

In 1991 Mount Pinatubo in the Philippines erupted to dramatic effect. As well as the vast quantities of ash that it rained down on the land around, it spurted a huge mass of aerosols twenty miles up into the air. This was so high that the aerosols lingered far above the weather systems that could otherwise have rained them back down to Earth. Over the next eighteen months they spread around the world, and global temperatures cooled by about 0.9°F. Dramatic eruptions like these cause aerosols to go high enough, and live long enough, to affect the climate. If such eruptions temporarily waned, that could cause the world to heat up.

It's a nice idea, but unfortunately it turns out to be wrong. There was a temporary lull in volcanic activity between about

1915 and 1956, but since then the world's volcanoes have been enthusiastically blowing their tops. In fact, they resumed their activity just before the warming took off.[5] If anything, volcanic aerosols have been keeping a slight check on the temperatures in recent years. To find the reason for the warming itself, we have to look elsewhere.

Spread the Sunlight Around Differently As Part of a Short-Term Natural Cycle

It's possible that we're just going through a warm phase of some perfectly normal cycle that spreads the heat out differently, sometimes warming, sometimes cooling Earth. We know that our planet has several of these cycles, most of which are not fully understood. Perhaps the most famous are the Milankovitch cycles. These are slight wobbles in the orbit of Earth around the sun, which affect how much sunlight arrives in the northern hemisphere in summertime on a timescale of 100,000 years or so.

Scientists believe that this is what triggers the planet to cycle in and out of ice ages. It works because the northern hemisphere contains most of Earth's land, and you need land to make glaciers. Slightly cooler northern hemisphere summers mean that snow from the previous winter can stay around on the ground. Gradually, the snow turns to ice and giant glaciers begin to form.

The change in the amount of sunlight in itself isn't large enough to explain the ice age. But other feedbacks quickly start to kick in. White ice reflects more sunlight than dark land, so Earth itself begins to cool. Then, as the planet cools, various other mechanisms start to suck carbon dioxide out of the air, which makes for even more cooling. The upshot is that a small shift in the amount of energy falling on one part of the planet causes a massive change in climate all over the world.

(Incidentally, this is why the temperature starts to drop a little in the ice core records before carbon dioxide levels change. Many skeptics argue that this time lag means that carbon dioxide can't ever be responsible for temperature changes. In fact, they're misreading the record. In ice ages, what happens first comes from outside. But it's only when this triggers changes in carbon dioxide that the real temperature rises and falls start to kick in. After that, the ice core record shows remarkably well how tightly changing carbon dioxide is coupled to changing temperature. Once the change in CO_2 has been triggered, the two march in impressive lockstep.)

The likeliest candidate for warming on the timescale of years, rather than millennia, would be the so-called El Niño/ Southern Oscillation. Roughly speaking, El Niño occurs when a warm pool of water that usually resides in Indonesia heads east toward the west coast of South America. Through a complex series of atmospheric connections, this process tends to warm the world a little. When the pool moves back west, the world correspondingly cools.

However, the warming effect of changing El Niño lasts only a few years. It also tends to have a very specific spatial effect; though the globe as a whole warms, the North and South Pacific tend to cool down. Our recent warming, by contrast, has been felt everywhere.

In fact, the same sorts of arguments apply to any form of internal cycle you can think of. As the IPCC report put it: "No known mode of internal variability leads to such widespread, near universal warming as has been observed in the past few decades."[6] Moreover, as we discussed in chapter 1, nature's temperature records show that we haven't seen warming like this for at least one thousand years, which makes it even less likely that we are on the up phase of some natural cycle.

Which leaves us with . . .

Trapping More Infrared Radiation As It Tries to Leave, Otherwise Known As the Greenhouse Effect

Chapter 1 describes the evidence that greenhouse gases affect temperature, and also the evidence that their concentrations have been rising. But there are additional reasons to believe that greenhouse gases are indeed the cause of the recent warming.

For one thing, unlike aerosols, which are rained out of the atmosphere relatively quickly, greenhouse gases are long-lived. Methane stays in the atmosphere some twelve years. Carbon dioxide remains in place for more than a century. Since both of these gases have plenty of time to spread right around the globe, their effect should be felt everywhere. And it is. The warming we have experienced since 1970 has shown up clearly on every inhabited continent.

Also, the heating effect from carbon dioxide tends to hug the ground. Normally, upper parts of the atmosphere soak up some of the outgoing heat and get a little warmer. But when carbon dioxide traps the heat below, like a blanket, the lowest part of the next layer up, the stratosphere, should begin to cool. This is one of the characteristic signatures of the greenhouse effect in action, and it's exactly what has been happening. Satellites and balloon-borne measurements show that the lower stratosphere has cooled by between 0.5°F and 1°F per decade since 1979, which is just what you'd expect if greenhouse gases were trapping radiation.[7]

The final proof that greenhouse gases really are the problem comes from atmospheric models of how the air works. Models have come a long way in the past few decades. The most sophisticated ones, called Global Climate Models,[8] or GCMs, use large, fast (and expensive) computers to calculate the behavior of the air. They envisage the atmosphere as adjoining towers of boxes, all of which obey the basic laws of physics. Air, heat, and moisture pass between the boxes, and radiation goes in and comes out. The air can also interact with the ocean, and in many cases with vegetation on Earth's surface.

GCMs have been criticized because it's not always easy to tell how realistic they are. Even if they do an excellent job of simulating the world we see, that could be because their operators have tweaked them to make sure they fit. Fortunately, though, we have plenty of information about how climate has changed in the past, often from the same sort of proxy measurements that we described earlier in chapter 1, such as ice cores, tree rings, and corals. So it's possible to test how good GCMs are by looking at how well they can reproduce the past. And, generally, the best ones do pretty well.

The models are not perfect. For instance, they still can't get down to the sort of regional detail we need to predict the events that will affect individual communities, such as storms and local shifts in rainfall patterns; they're not yet very good at simulating clouds; aerosols provide a particular challenge, and for this and other reasons different models still give a spread of answers about exactly how much temperature rise will come from a given rise in carbon dioxide. We address this further in chapter 6.

But there is one thing that models can do that nothing else can manage, and that's experiments involving the entire planet. Normally, if scientists want to establish the cause of a particular effect, they try an experiment in different ways. What happens if I add this? Let's try again taking away that. But with the climate problem they are stymied—there is, after all, only one Earth.

Models can help to fill this gap. You can let a model run with different inputs—only solar activity, say, and volcanoes, or only greenhouse gases—and then see what the results look like. Comparing that to the real world gives us clues about what was really to blame.

And when the GCMs do that, they all come to the same conclusion. There is no way to explain the warming of the past few decades unless you include the rise in greenhouse gases. But when you do add the gases, you see exactly what happened in the real world. The same story applies individually to every inhabited continent on Earth. Each one has seen dramatic warming in the past few decades, and in every case the models can account for the warming only if greenhouse gases are added to the mix.

In fact, the models generally do a good job of explaining all the changes that have taken place in the twentieth century, including the fact that temperatures apparently dropped a little during the middle part of it—something of which climate change skeptics like to make great play. It turns out that the cooling came from something we've already discussed in this chapter: aerosols. They did in fact have a marked effect on the temperature of the twentieth century. It was anything but natural, however. Burning dirty coal produces plenty of sulfur-containing aerosols, and researchers now think that these were responsible for the slight cooling that took place between about 1940 and the late 1960s.

Some darker forms of aerosols formed from burning can have the opposite effect. Black or brown soot particles cause warming because their dark color means that they soak up sunlight, and thus heat the air around them. For instance, the brown clouds of smog that hover over parts of the Indian subcontinent help to explain why parts of this region are heating up much faster than the global average, which could in turn explain why the Tibetan glaciers are receding at such an alarming rate.[9] But sulfate aerosols have a cooling effect, and they dominated during the middle of the twentieth century.

The early part of that century was hit by the cooperative effects of increased sunlight, reduced volcanic activity, and increased greenhouse gases, counterbalanced a little by the aerosols from dirty coal. The overall effect was one of warming. But as the sun's

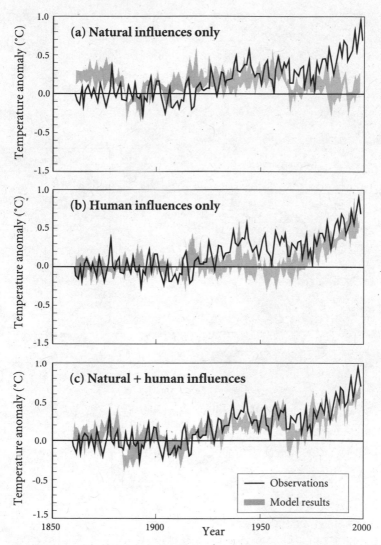

Temperature changes over the past 150 years in °C compared to the 1961–90 average. The models cannot reproduce this unless they incorporate both natural and human influences. (Source: IPCC)

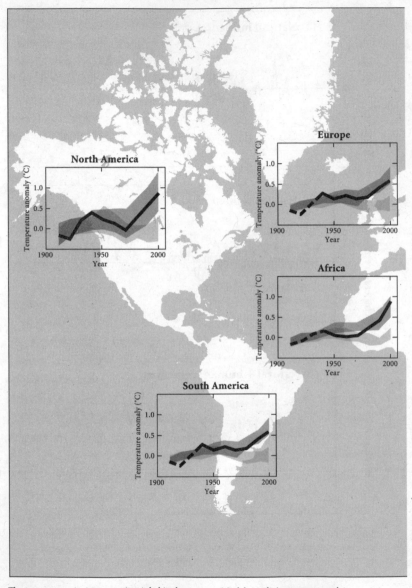

The temperature is rising on every inhabited continent. Model simulations cannot explain this rise unless they incorporate the effects of human-induced changes in greenhouse gases. (Source: IPCC)

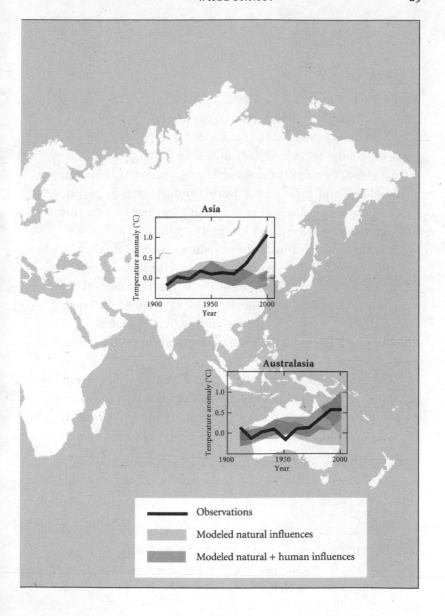

Asia

Australasia

Observations

Modeled natural influences

Modeled natural + human influences

activity waned, aerosols began to win this tug-of-war, and for a few years Earth cooled slightly.[10]

One major clue that this is correct comes from the fact that the cooling was seen only in the northern hemisphere. The aerosols from fires and factory chimneys are not thrown high into the atmosphere like the ones from Mount Pinatubo, and so they tend to be rained back out again after a week or so. Thus they don't have time to spread, and they have to linger where they were formed. Because the northern hemisphere has most of the world's land, and most of the world's industrialized countries, that's where the aerosols were. And that's why only the northern hemisphere witnessed this midcentury cooling.

But then many industrialized nations cleaned up their act. Nobody wanted to live in choking, filthy cities, and a series of Clean Air Acts were passed. Ironically, in cleaning away the obvious signs of pollution in our air, we lost some of the cooling effects that these particles were giving us. And we left the invisible and much more deadly pollutants, greenhouse gases, free to do their worst.

The models show this beautifully. When they put together solar warming and greenhouse gases, the early part of the century in the model warms exactly as it did in the real world. The aerosols gradually take over in the middle of the century, and then the greenhouse gases win out in the end.[11]

A few other apparent holes in the greenhouse gas argument were eagerly seized on by skeptics. But in the last few years they have all been resolved.

For instance, though ground-based measurements showed clearly that Earth was warming, satellites seemed to suggest that just a little higher up, in the mid- to upper troposphere, temperatures were not moving. But this turned out to be because of straightforward errors in how the satellite data were analyzed. Proper analysis reveals that upper parts of the troposphere are indeed warming at a very similar rate to the ground.

Also, some people suggested that carbon dioxide would not increase the amount of water vapor in the air, because the upper part of the troposphere would dry out as the lower part moistened. Without the power of the extra water vapor to boost the carbon dioxide's greenhouse effect, they suggested, it wouldn't cause enough warming to matter. However, satellite measurements now show that the upper troposphere is indeed getting wetter, exactly as predicted.[12]

All of this evidence points to the same thing. The recent heating up of planet Earth has carbon dioxide's fingerprints all over it. Or, to put it more succinctly, and to answer the question posed in the title of this chapter: Wedunnit. Human activity is to blame for the rise in temperature over recent decades, and will be responsible for more changes in the future. There are plenty of areas for debate in the global warming story, but this is not one of them. Anybody tells you differently either has a vested interest in ignoring the scientific arguments or is a fool.

Until now, we have focused on the changes in temperature that planet Earth has experienced in the past few decades. But these changes have also affected many other aspects of the climate, with corresponding trouble for its inhabitants. We'll talk about this in the next chapter.

3

FEELING THE HEAT

So far we have experienced only a relatively small amount of global warming—about three-quarters of a degree centigrade. But that has still been enough to throw a substantial wrench into nature's workings. It's not just that animals and plants can't take the heat, and simply keel over and die. Rather, their lives were carefully tuned to a climate that no longer exists.

The essence of the problem is timing. Animals use very specific cues to make decisions about when to emerge from hibernation, have sex, or set out on a journey that will take them to the other side of the world. And cues that used to be ultrareliable are now beginning to fall short. Around the world, spring is coming progressively earlier, and autumn ever later. Though some animals and plants are managing to cope, the natural world is like an orchestra. If the timing is off, you don't get beautiful harmonies—you get a cacophony.

Take the yellow-bellied marmot, a ground squirrel that lives in the mountains of the western United States and Canada. Each year, these creatures decide when to emerge from their underground dens by testing the air temperature. Warm air ought to mean that the snow will soon be melting and food is on its way. But of late the warm air has come much earlier, when snow is still thick on the ground. Marmots have started emerging more than a month before they should, and are vulnerable to both

lack of food and the telltale tracks in the snow that lead predators to their door.[1]

The Edith's checkerspot butterfly uses snow cover, rather
than temperature, to decide when to emerge from its chrysalis
and make its dash for reproduction. But in its Rocky Mountain
habitat, snow has been failing, tempting the butterflies out in
April instead of May or June. In one year there were no plants
to feed them, and the butterflies' bodies littered the surrounding hillsides with a bright orange carpet of fresh, soft wings. After several more abortive attempts to reproduce, they have now
vanished from this part of the world.

Changes in rainfall are also having their insidious effect.
Generally speaking, northern parts of the world have been getting wetter while the tropics have been drying out. More rain
has come to the eastern states of the Americas, northern Europe, and northern and central Asia, whereas North Africa,
southern Africa, the Mediterranean, and parts of southern Asia
have all begun to lose water. Even in places where the annual
rainfall has stayed more or less the same, the pattern has begun
to change, with more intense rainstorms and longer, drier periods in between.

The lemurs of Madagascar are a case in point when it comes
to the subtlety of climate effects. These remarkable primates range
from the minuscule mouse lemur, which is small enough to curl
up in the palm of your hand, to the indri, which is famous for its
haunting song and is about the size of (and looks rather like) a
four-year-old child dressed in a baggy panda suit. A host of other
species fit in between these two size extremes and each has its
own individual calendar for mating. But what's really remarkable
is that they all aim to wean their offspring at more or less the same
time—when the fruits of the forest are at their most plentiful.

Lemurs, like many other animals, know the date from the
length of time they see daylight. This is the one constant in a

shifting world. At any point on Earth's surface, the length of daylight is a function of where we are in our orbit around the sun; it doesn't depend on temperature or rainfall or other capricious measures. Thus, the amount of daylight tells lemurs the date as surely as if they had a red ring scrawled on a calendar.

This approach works beautifully as long as the fruits play along. But if rainfall patterns change, even though the calendar may be right the availability of food can be very wrong. And there are already signs that this could be happening in Madagascar. The effect may be subtle: a little less nourishment, fewer young that survive the year, and a group that is just that much smaller. But with animals that have already been driven to the edge of extinction by the chain saws that have destroyed most of their natural habitat, the subtle effects of climate change could easily be enough to tip them over.

The pied flycatcher has also begun to lose faith in its internal calendar. Each year these birds make an incredible journey of three thousand miles from western Africa to their breeding grounds in Holland. Their starting date is strictly marked, and for the past twenty years they have been arriving at the same time. But local temperatures in Holland have begun to change. Spring is coming earlier to Europe, and the caterpillars that the flycatchers' chicks are supposed to eat are now past their peak. If the flycatchers don't evolve a new starting date, they will die.

Some species are managing to adapt.[2] However, human-induced climate change is happening much faster than it does from natural causes. The result is that the creatures that can evolve with it tend to be the ones that are quickest on their evolutionary feet—which means flies, mosquitoes, and other insects with short reproduction times rather than the bigger, more photogenic animals.

There are many other signs that the natural world is already feeling the heat. The tree lines are marching farther toward the poles and up the sides of mountains. Animals are following their plant food up into the hills. Carefully sited national parks, designed to protect key habitats, are looking increasingly vulnerable, and conservationists are beginning to realize that climate change could be the final straw for many of the world's hard-pressed species.

No single species has yet earned the dubious honor of being the first undisputed victim of climate change. The best candidate to date has been Costa Rica's exquisite golden toad, which disappeared forever in the mid-1980s from a rain forest that was apparently untouched by humans. At first, researchers suggested that the toad had died out as changing temperatures removed the mist that wreathed its cloud-forest home. But it later seemed more likely that disease had played a significant part in the loss of this and a large number of other Costa Rican frogs in the 1980s and 1990s. Still, in 2006 the original researchers published a paper making a very convincing case that climate change had created the perfect conditions for the disease to spread.[3]

Though many feel sad at the thought that the golden toad has hopped its last on Earth, it's easy to think that, with so many species on the planet, we can probably spare a few without suffering much worse than a twinge of conscience. But that doesn't give enough credit to the potential scale of the losses. The IPCC report predicts that between one-fifth and one-third of all species on Earth could be at risk of extinction by the end of this century,[4] and other researchers have even suggested it could be more than one-half. That is enough to qualify for a mass extinction on the scale of the dinosaurs.

For now, what's perhaps more alarming is that global warming isn't only affecting individual species; it's changing entire ecosystems. The two most dramatic examples are coral reefs that are turning white, and the north polar ice cap, which is turning black.

The Case of the Disappearing Pole

Antarctica is an ice-covered continent surrounded by water, and the ice on its land can be two miles thick. The Arctic, on the other hand, is an ocean surrounded by a ring of land. The ice at the North Pole floats on seawater and is only a few feet thick. Because of this, the ice there is much easier to melt.

Over the past thirty years, as temperatures have begun to soar, the area of summer sea ice has shrunk by nearly 8 percent per decade.[5] It's shocking to watch a satellite sequence of the disappearing ice. Each year, the area of whiteness shrinks perceptibly, to be replaced by dark ocean.

The past four years have seen record low areas of sea ice, with 2005 providing what one researcher calls the "exclamation point" in the record. That year had 20 percent less ice cover than the average, a loss of about five hundred thousand square miles.[6] And 2007 was even worse. Meanwhile, springtime temperatures began rising throughout the Arctic basin in the 1990s,[7] and many parts of the Arctic have begun to experience unprecedented heat waves.

Models predict that if greenhouse emissions continue the way they are going, by the end of the century the summer Arctic will be completely ice free.[8] A planet that used to boast two polar ice caps will then have only one.

Already, the ecosystem is beginning to change. For instance, there are signs that the loss has begun to hurt polar bears, the poster child of the global warming story. Polar bears are specialist predators that feed almost exclusively on seals. The seals, in turn, breed on sea ice. So access to sea ice is vital for the bears, especially for the females that emerge from land-based dens with their cubs in the summer after a fast that has lasted six or seven months. If, on emerging, they face nothing but dark ocean water, they must either swim or starve. One group in western

Hudson Bay has already declined from 1,200 bears in 1987 to fewer than 950 in 2004.[9]

Losing the sea ice is bad not just for the larger, more picturesque members of the Arctic ecosystem, but for many of their smaller prey—right down to the ice-dwelling phytoplankton that make up a considerable proportion of the food supply.[10] It's not yet clear exactly how the Arctic species will shift to adapt to this new, wetter Arctic, but the entire ecosystem will certainly change beyond recognition.

Bleaching Corals

Coral reefs harbor so many colorful species that they're often called the "rain forests" of the ocean. In spite of their hard appearance, they are actually jellylike creatures that quiver inside a calcium home of their own making. Sharing their shell are house guests in the form of algae, which provide the corals with food. Normally this is an arrangement that suits both parties, but when the coral becomes stressed, the first thing to go is the tenant. The algae are unceremoniously ejected, and the coral turns white. This is known as coral bleaching, and it occurs in particular if the seawater grows too hot.

That's what happened in 1998, which was the joint warmest year in the entire historical record. Bleaching began in the eastern Pacific, in French Polynesia, before it headed out west. For some reason it skipped over many of the western South Pacific islands, but it wrought havoc on the Great Barrier Reef. The damage was so dramatic there that the gigantic scarring was visible from space. The bleaching went on to the Indian Ocean, where it destroyed 90 percent of the corals in the Maldives, then to Africa and the Caribbean. The sheer global scale of the attack took researchers by surprise. Nobody had seen anything like it.

The bleaching returned in 2000, this time wrecking large swathes of Fijian corals. In 2002 the Great Barrier Reef was hit again, and in 2005 the Caribbean suffered even worse losses than it had in 1998.

Corals can survive bleaching episodes and retrieve their algae when the temperature cools. But if the warming is too great, or lasts too long, the coral dies. That's not to say it goes extinct. Reefs can be recolonized, or corals can set up new colonies elsewhere. They play the numbers game. So many coral larvae are released at each spawning that millions of individuals can die without the species disappearing altogether. They can also migrate to places with more appropriate temperatures and some may be able to adapt quickly enough, evolving a way to tolerate the higher temperatures and greater acidity. But as a result of the bleaching, many of the world's existing coral reefs are already dying. And as ocean temperatures continue to rise, the bleaching will only get worse.

Increasing carbon dioxide has another side effect that will also be bad news for corals. To date, the oceans have soaked up around half of the carbon dioxide emissions from burning fossil fuels, making cement, and land-use changes. That's just as well; otherwise there would be even more carbon dioxide in the air and hence even more warming than we have seen so far. But the benefit has come at a price: All that additional carbon dioxide is gradually acidifying the ocean.

This seems hard to believe and, indeed, many scientists initially discounted the possibility. The oceans, after all, are vast and are also expert at neutralizing any material that threatens to acidify their waters. However, the carbon dioxide is arriving too fast. It's overwhelming the ocean's natural capacity to compensate. A report produced by Britain's Royal Society in 2005 estimated that the world's oceans had already increased their acidity by 0.1 units, which translates to an increase in the ions that cause acidity of some 30 percent.[11] And one study calculated that unfettered carbon dioxide increases over the next few centuries could make the oceans more acid than they have been for three hundred million years.[12]

Nobody yet knows whether this is having an impact on the world's sea creatures, mainly because it is hard to measure. But most agree that the effects will be felt soon, if, indeed, they are not already with us. According to the Royal Society report, animals with a high metabolism, such as squid, are likely to suffer in more acidic waters. But the real danger is to any animal that makes itself a shell, or skeleton, out of the calcium carbonate (the same stuff as common chalk) dissolved in seawater. The more acidic the seawater, the harder it is to make this shell, and in the extreme the shells already made will begin to dissolve.

This danger applies to creatures that span a wide part of the food chain—from tiny plankton and pteropods, which feed cod, salmon, and whales, to mussels, conch, and sea urchins. It applies particularly to corals.

Sixty-five million years ago a meteor the size of New York slammed into Earth. The environmental chaos that ensued is widely believed to have led to the extinction of the dinosaurs. But it also had a less well-known effect. According to Ken Caldeira at the Carnegie Institution of Washington in California, the meteor also threw up vast amounts of sulfur, which then rained down on the ocean as sulfuric acid. The upper ocean became acidified for a brief moment, perhaps only one or two years. But that was enough. More or less every sea creature that built shells or skeletons out of calcium carbonate became either rare or extinct. A handful of corals must have survived, or we would not still have them on Earth today. But they were nonetheless too scarce to leave their imprint; they did not reappear in the fossil record for a full two million years.[13]

In all, the IPCC report states that more than 29,000 observational data sets in seventy-five different studies show significant changes in physical and biological systems around the world,[14] 90 percent of which go in the direction you would expect because of warming. What's more, the regions where the changes are happening coincide with the regions where the most warming has been seen. This can't be simply the result of natural variability. The message from the natural world is that climate change is already here.

This is not merely bad news for lovers of wildlife. Our ecosystems work for us in ways that we're so used to that—in Joni Mitchell's memorable words—"we won't know what we've got till it's gone."[15] These so-called ecosystem services provide more than an aesthetic backdrop for a walk in the woods. They also give us food and natural materials and do a host of other jobs that include soaking up our pollution, delivering water where we need it, and holding it back where we don't.

Though it is always hard to attribute any one occurrence to climate change, three recent dramatic events have helped to throw into stark relief the human dangers of interfering with the climate.

Climate Wars

Until a few decades ago few Westerners would have heard of Darfur, a region of western Sudan just south of the Sahara. This was an essentially peaceful region where nomadic Arab herders grazed their animals amicably on the land of African farmers. Now, of course, Darfur is a place of almost unimaginable horror. Something happened to set those erstwhile peaceful neighbors tearing out each other's throats. More than 2 million people have been displaced, their homes destroyed and their lives ruined. At least 400,000 civilians have been killed.

The origins of the conflict are complex, but they were triggered by an ongoing drought that took its first terrible hold in Sudan and the Horn of Africa in the mid-1980s. Life that was already on the edge was about to tip over it. People and cattle died in horrifying numbers, and the scarce resources of water and fertile land that remained became a breeding ground for conflict. In the late 1980s the increasingly desperate Arab herders began to attack their farming neighbors. The result we know only too well.

At first, it was assumed that the drought that triggered this war over resources had been caused by environmental degradation. Westerners tutted that irresponsible overcropping and overgrazing had turned the land into a dust bowl. But more recent models have shown that improper land wasn't the cause.

Instead, models suggest that the reason for the drought is tied to a cooling in the northern Atlantic relative to the southern. This in turn changed the location of a set of air currents known to scientists as the Intertropical Convergence Zone, and to the farmers and herders of Darfur as the bringer of rain.[16]

The picture now becomes muddier. It is possible that cooling of the northern Atlantic had natural causes. But it's also possible that it happened because of the partner-in-crime of carbon dioxide: sulfate aerosols. As we explained in chapter 2, these arise, among other reasons, from burning dirty coal. Because they reflect sunlight directly back out to space, they are engaged in a perpetual tug-of-war with carbon dioxide, causing cooling that partly balances the greenhouse heating. Although many of these aerosols have been cleaned up in the past few decades, they still billow from power stations and factory chimneys in enough numbers to cast a pall over the oceans that abut the industrial centers of the world.

And the key effect with regard to Darfur was a purely regional one. One model suggests that over the North Atlantic clouds of aerosols may have bounced a little sunlight back to space, which cooled the local sea surface, which in turn sent the Intertropical Convergence Zone southward and robbed the Sahel of its rain.[17]

Of course what followed the drought was politics—notably the neglect of the region and then the unleashing of power-hungry warlords by Sudan's central government at Khartoum. But the violent, horrifying conflict might never have taken hold had the burning of fossil fuels not laid the groundwork. As one writer put it: "In decades to come, Darfur may be seen as one of the first true climate change wars."[18]

Whether or not this is true (and the models have yet to agree among themselves), Darfur remains a dramatic example of how a small shift in climate in just one region can have dramatic and horrifying human consequences.

Katrina

The same sort of argument applies to another recent disaster. On the morning of August 29, 2005, Hurricane Katrina struck southeastern Louisiana. It was the sixth strongest Atlantic hurricane ever recorded, though it weakened considerably before it hit land. It was also one of the deadliest Atlantic hurricanes. Katrina caused $80 billion in damage and took more than 1,800 lives. It also, infamously, turned New Orleans—a major city in the world's richest country—into a hellhole, a place where bloated bodies floated through the streets and where lawlessness and disease were rampant.

Though many people have suggested that Katrina was a direct consequence of climate change, we don't know whether this is true. Think of a dice loaded so that it is twice as likely to throw a six. Half of the sixes that you throw afterward would be "natural" ones, but the other half would come from the change that had been deliberately made. We don't know—can't know—

whether Katrina was the result of a "natural" throw of the climate dice or of one of the loaded ones. What we can say is that, thanks to climate change, more Katrinas are likely to be on their way.

Warming is unlikely to make a difference to the number of hurricanes around the world,[19] but it is very likely to make them more intense. The energy that feeds hurricanes comes from warm seawater, which is why they form only in the tropics, and then only in the hottest months. Global warming is already causing the sea surface to heat up,[20] and that in turn is increasing the hurricanes' chances of becoming truly destructive. More intense hurricanes also have more water to work with, as rising sea levels give them ever greater reach.

Sea temperature isn't the only influence on the ultimate power of a hurricane. If wind speeds change at different heights above the water surface, this can slice up the baby hurricanes before they are able to grow big and strong. However, historical evidence suggests that rising temperatures are already having an effect. Over the past thirty years, hurricanes have become roughly 70 percent more energetic.[21] Since 1970 the numbers of category 4 and 5 storms—the strongest and most destructive—have dramatically increased, with the areas worst hit being the North Pacific, Indian, and Southwest Pacific oceans. Hurricanes may even be expanding their historical boundaries: In March 2004 the first-ever tropical cyclone in the South Atlantic smashed into the coast of Brazil. It took forecasters so completely by surprise that they hadn't even given it a name.

Heat Wave

In the summer of 2003 an intense wave of heat spread across Europe. While the northernmost countries basked in the unexpected warmth, the rest of Europe fried. Wilting crops cost farmers more than $12 billion. Forest fires in Portugal caused another $1.6 billion worth of damage.[22] And people died. The statistics for numbers of deaths per month across the continent show an enormous spike in the first few weeks of August, the hottest month. They occurred among people in every age group from forty-five upward. Most were already vulnerable: older people, asthmatics, people who literally couldn't stand the heat. And yet they were not already close to death. If they had been, the abrupt rise in deaths in August would have been followed by a drop over the following few months as the statistics caught up. But this didn't happen. Instead, at least thirty-five thousand, and perhaps as many as fifty-two thousand, people died significantly before their time.[23]

So was this just a "natural" blip or was global warming to blame? In this case, unlike the two previous ones, the evidence for the involvement of climate change is compelling. In 2004 several British researchers ran one of the world's top climate models with only natural variations in solar and volcanic activity and then ran it again with greenhouse gases added. They then calculated how much the greenhouse gases had increased the risk of a summer at the 2003 level. Their results showed that global warming had made such a summer much more likely than before.[24]

One way to understand this is to think about the difference between climate and weather. Weather is what we experience month to month and year to year. It fluctuates constantly, and a graph of the changes in annual temperatures in central Europe over time shows a forest of small spikes, some going up, some down. Climate is the average of these fluctuations over time, a line drawn through the middle of the spikes.

Of late, our climate has been warming, so this steady average line drawn through the spikes that mark our annual temperature has been rising. This means that any upward spike that is otherwise a fairly ordinary departure from the average is now

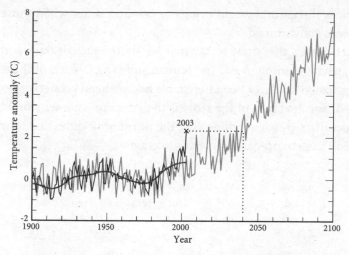

Actual and smooth summer temperatures in Europe in °C (black lines) compared to the 1961–90 average, ending with the 2003 European heat wave (cross). According to the models (gray line), by 2040 the value of 2003 will have become typical. (Source: P. A. Stott, D. A. Stone, and M. R. Allen, "Human contribution to the European heat wave of 2003," *Nature*, vol. 432, pp. 610–14, December 2, 2004)

starting from a higher baseline. It peaks at a much higher temperature than it would otherwise have reached.

In this way, fully half the excess temperature of 2003 came from the rising baseline. The rest came from what would otherwise have been a high, but not especially unusual, temperature spike. The baseline has risen so much that the average summer temperature in central Europe is now close to the value for the hottest summer of the twentieth century. In the next few decades, the baseline looks likely to rise to match the temperatures of 2003. Then, when a hot summer spike comes along, the effect will be truly searing.

These are just a few examples of ways that relatively small changes in climate have already caused massive human suffering. Like it or not, insulated in cities or out on the prairie, or the steppe, or the savannah, we all depend fundamentally on the services the natural world brings to us—and suffer when these services are withheld.

However, that's not to say that we should just sit back and take what's coming to us. The human suffering in each of these cases would not have been inevitable had we been properly prepared. The next part of the story is to figure out what we should expect from climate change over the next few decades and how we will have to prepare ourselves to cope.

4

IN THE PIPELINE

Change Is Inevitable

Most people have now realized that climate change is upon us. If pushed, most would probably also say that if we don't do something to change the way we live, things are likely to get worse. But few seem to have noticed one of the most important points to emerge from the last few years of scientific projections. All the evidence suggests that the world will experience significant and potentially highly dangerous changes in climate over the next few decades *no matter what we do now.*

That's because the ocean has a built-in lag. It takes time to heat up, which is why the nicest time to swim is often the end of the summer rather than the middle. The same principle holds for global warming, but on a longer timescale: Because the oceans gradually soak up heat generated by the extra greenhouse gases, the full effect won't be felt for decades to centuries.

This means that whatever we do now to change our carbon habits will take several decades to have any effect. In other words, according to our most sophisticated models, the next twenty to thirty years will be more or less the same whether we quickly kick the carbon habit or continue burning as many fossil fuels as we can.[1] Whatever we do today to reduce emissions will matter for our children's generation and beyond, but not for our own. The problem of climate change is one of legacy.

We'll talk more about this later. For now, the important point is that a certain amount of climate change is already in the pipeline. Though some of the details are sketchy, scientists have a reasonable idea of what the next few decades will hold. Note that the descriptions that follow assume we do nothing to prepare for the changes that it's now too late to stop. Later in the chapter we will explain some of the ways in which we can learn to adapt.

Storms

As we mentioned in chapter 3, hurricanes will probably not increase in number, but are likely to get stronger and last longer. Outside the tropics the picture is more complicated. Storms there are much larger and more sprawling than hurricanes. They deliver unpredictable weather, but also essential rain to large swathes of the middle latitudes in both hemispheres. Whether they change in number or intensity depends on a complex set of variables including temperature, moisture levels, and how quickly the temperature changes as you go from the warm tropics to the cold poles. As yet, the models don't agree. However, one thing that seems clear in all the models is that the storm tracks—the upper-atmosphere "guide rails" around which the storms roll—are likely to shift poleward, which would mean less rain in southern Australia, the northern United States, and parts of southern Europe. Most models also predict a strengthening of the fierce westerly winds in the Atlantic, which would mean stronger winter storms hitting northern Europe.[2]

Water

There will be less snowfall, fewer glaciers, more variable rainfall patterns, and more intense individual rainstorms. By the middle of the century, high latitudes and some wet tropical areas will become up to 40 percent wetter, with more water flow from rivers and generally more water availability. But some regions in the middle latitudes (such as southern Europe and the southwestern United States) and in the dry tropics will become up to 30 percent drier. Disappearing glaciers and snow cover will reduce the meltwater that normally flows from mountain ranges, which will affect more than a billion people.

In West Africa, twenty-five watercourses cross the boundaries of seventeen different countries, and as the region dries, there will be an increasing danger of the sort of conflict that has devastated Darfur. To the north, ten countries share the resources of the Nile River. While the population in these nations is set to burgeon, many climate models are predicting that the waters will recede.[3] Even in places where the annual rainfall drops, individual storms are expected to dump more water in intense bursts. Nearly one and a half billion people live in river basins that are more likely to flood as a result.

Heat Waves

In the next few decades, heat waves such as the one that caused devastation in Europe in 2003 will become increasingly common. By the 2040s, for instance, at least half the European summers are likely to be just as hot.[4] Many other regions are likely to suffer the consequences of dry, hot summers, especially increasing wildfires such as the ones that scarred mainland Greece in 2007. This will, however, come with a climate upside: There are also likely to be fewer seriously cold winters. Since deaths from the cold usually outnumber those from heat by a considerable margin, this means there could be an overall decrease in the number of temperature-related deaths. However, as the authors of one study put it, "This offers little reassurance for those affected by the heat."[5]

Monsoons

There is considerable uncertainty about which way the monsoons will go. A combination of the local cooling effect of aerosols over the Atlantic and Indian oceans and the possible weakening of the ocean circulation suggests that monsoons may become drier, which would be potentially catastrophic for regions whose livelihoods are sensitively dependent on the annual rains.

On the other hand, many models predict that the Asian monsoon is likely to get gradually wetter. That sounds like good news, as it would bring more water to roughly two billion people in East and South Asia. However, this is likely to fall in even more intense bursts than normal, which could mean severe flooding. In August 2005 Mumbai experienced a record-breaking rainstorm. Three feet of rain fell on the city in just hours. The resulting floods closed schools, banks, the stock exchange, and the airport. More than a thousand people died.[6]

Food and Forests

The good news is that global food production is set to increase over the next few decades. The devil, however, is in the regional detail. For middle to high latitudes, crop productivity is likely to go up. (Though if the temperature creeps up more than 3.5–5°F, productivity tumbles even there.) By contrast, at lower latitudes, especially in tropical regions, even the small rise in temperature predicted for the next few decades will send crop production tumbling. That, of course, covers most of Africa and parts of Asia whose populations emit only a tiny amount of carbon dioxide per person relative to Western countries. It's a nasty irony that the people least responsible for the problem will also be the ones to suffer first, and most.

Rising Sea Levels

As the oceans warm they expand. This effect, coupled with melting glaciers, is already raising sea levels all over the world. For the twentieth century, the rate of rise was about eight hundredths of an inch per year, but of late it has risen to more than one-tenth of an inch and counting. Most researchers think that the rate will accelerate as warming temperatures eat farther into the great ice sheets of Greenland and Antarctica. But even over the next few decades, the current rate is enough to produce a rise of up to four inches.

That might not sound like much, but it's also important to realize that the rise isn't uniform. The ocean is surprisingly lumpy; currents and the spinning planet send water sloshing from one ocean basin to the next. Thus, in some places the rise in sea level will be greater, and in others sea level will even fall. (Despite what skeptics say, falling sea levels in some regions are a perfectly normal consequence of the way the oceans behave, and emphatically not a counterindicator of climate change.)

The rising ocean will bring erosion and flooding to many coastal regions, as well as turning coastal wetlands into salt marshes. More than three billion people, roughly half the world's population, live within 125 miles of a coast. Many of these are close enough to suffer from what will be increasingly deadly storm surges that bring the sea flooding inland. And many, too, live in developing countries where they have the least financial resources to adapt and are the most dependent on local food and water supplies. But the rising sea does not pick and choose, and it will equally affect the big coastal cities of industrialized countries, such as London, New York, Tokyo, and Sydney.

To some extent this is already happening. The Thames Barrier stretches across the mouth of the Thames River and can

	WATER SHORTAGE	FOOD PRODUCTION	DISEASE
Africa	From 75 million to 250 million people at risk by 2020s.	Reduced yields across the continent, in some countries as much as 50 percent by 2020. Hundreds of millions at increased risk of famine.	Malaria likely to reduce range in southern Africa but to begin to extend range to eastern highlands.
Asia	From 120 million to 1.2 billion people at risk by 2020s.	A 2.5 to 10 percent decrease in crop yield in 2020s. Up to 49 million at risk of hunger.	Flooding and high temperatures increase risk of infectious diseases, especially cholera and typhoid.
Australia, New Zealand, and Small Islands	Reduced water runoff in most of eastern and southwest Australia, with a fall of up to 20 percent in the southeast by 2030. Significantly less snow cover in the southeast.	Enhanced growing conditions for much of New Zealand and wetter parts of Southern Australia. Possible reduced yields in parts of eastern New Zealand and Southern/Eastern Australia that are far from large rivers due to a combination of drought and fire. Risk to fish supply from corals around small tropical islands.	An additional 0.1–0.3 million exposed to the risk of dengue fever by 2020.

Anticipated climate changes over the next few decades, broken down by region. (Source: IPCC)

FLOODING	HEAT WAVES AND FIRE	CORALS AND ICE
High risk to large coastal cities such as Lagos and Alexandria.	Combination of drying and temperature rise brings increased threat of burning to many forests.	Red Sea and East African corals at risk of bleaching.
An estimated 2.5 million people flooded by 2050, mostly in the Ganges-Brahmaputra and Mekong megadeltas. Many more at risk from more intense typhoons, rising sea level, and storm surge. More variable monsoon might bring flash flooding.	Increased mortality from severe heat waves for the southern and eastern parts of the continent, and possibly also Siberia. Reduced mortality from serious cold events.	Very substantial or complete loss of Himalayan glaciers by 2035, together with loss of 30 percent of coral reefs.
Rising sea level and more intense cyclones increase risk of coastal flooding in Australia and New Zealand. Very substantial risk of increased flooding in low-lying small islands, as well as encroaching salt water into coastal fields.	A 4 to 25 percent increase in frequency of fire danger days in southeast Australia by 2020. Heat-related deaths to double in Adelaide, Melbourne, Perth, Sydney, and Brisbane by 2020.	A 58 to 81 percent of the Great Barrier Reef bleached every year by 2030. Also substantial bleaching of corals around small islands.

	WATER SHORTAGE	FOOD PRODUCTION	DISEASE
Europe	In Southern Europe surface runoff decreases by 0 to 23 percent by the 2020s. Longer droughts and significantly increased fire risks, especially in the Mediterranean.	Higher crop yields for northern Europe, but reduced yields for the south, especially in the Mediterranean, southwest Balkans, and the south of European Russia.	No significant change expected in infectious diseases.
Central and South America	From 7 million to 77 million people at risk by the 2020s.	Soybean yields could rise, but rice yields are likely to fall. Wheat and corn hard to predict but overall 5 million people at increased risk of hunger by 2020.	Possible reduction of malarial risk in Amazonia and Central America, but range could begin to extend beyond previous southern limit.
North America and polar regions	About 41 percent of water supply to Southern California vulnerable to loss of snowpack from Sierra Nevada and Colorado River basin.	Likely increases in crop yield of 5 to 20 percent over the next few decades, but crops currently near climate threshold, e.g., wine grapes in California, likely to suffer falls in yield, quality, or both.	Boundary for tickborne Lyme disease likely to shift 200 km northward by the 2020s.

FLOODING	HEAT WAVES AND FIRE	CORALS AND ICE
Increased storm intensity in the North Atlantic to 2030. Higher sea level, more severe winter storms, and more intense rainfall events threaten coastal and inland cities, especially in the north, where runoff is also projected to increase.	By 2030 heat waves such as the one in 2003 that killed more than 30,000 people will become the rule rather than the exception. There will, however, be reduced deaths from serious cold events.	Substantial loss of alpine glaciers; decrease in area of seasonal snow cover, especially at lower elevations.
Danger of flooding in low-lying areas, especially El Salvador, Guyana, and Argentine coast. More intense hurricanes in Caribbean.	More frequent wildfires in much of South America, including the Amazon rain forest, and in parts of Central America.	Substantial or complete melting of most tropical glaciers. Severe bleaching of Caribbean corals.
Coastal cities increasingly vulnerable to flooding, especially in the Gulf of Mexico, as tropical storms grow more intense.	Storm tracks moving north will bring droughts and fires to parts of the northern and western U.S., where average area burned each year has already almost doubled. Also increased fire risk in southern Canada.	Very substantial or complete melting of ice in Glacier National Park by 2030. Considerable loss of glaciers from Greenland, Alaska, northern Canada, and the Antarctic peninsula. Very substantial loss of Arctic sea ice by 2030.

be raised to protect the city of London from flooding. When it began operating in the early 1980s, it was used on average once every three years. Now it is being deployed up to six times a year.

Corals

Within the next few decades, the world's coral reefs are going to be hit hard by the double whammy of a more acidic ocean and rising sea temperatures. Although individual coral species are unlikely to become extinct, many existing coral reefs will bleach and die, which will in turn affect the people who rely on the fish they shelter for subsistence and tourist income.

Development

Many of the variations that climate change will bring over the next few decades will exacerbate the development problems already faced by large parts of the world. For instance, even in today's world eight hundred million people are undernourished, and the projected drop in food production in low-latitude areas will only make this worse. Similarly, about two billion people currently lack access to clean water. The rural poor depend more than anyone on ecosystems for their livelihoods. Reduced water flow in some areas may make energy supplies from hydropower less reliable. And, of course, there is the danger that competition for dwindling resources will exacerbate problems of local, national, and regional security.

Learning to Adapt

Three miles above sea level, high on the Tibetan Plateau, lies a lake called the Tsho Rolpa. Warmer temperatures have accelerated the melting of the region's glaciers that feed this lake, and between the 1950s and the 1990s the Tsho Rolpa's area grew by more than a factor of seven.[7] By then, a billion square feet of water were straining at a flimsy moraine dam, threatening a hydropower plant downstream as well as many villages. Therefore, the Nepalese government cut a channel into the moraine wall and installed a gate to control the release of water and lower the lake level. The construction project took four years and cost $3.2 million. Meanwhile, in case of flooding, an early-warning system was established for nineteen villages downstream. The villagers themselves helped to design the scheme and carry out periodic drills to ensure that everyone knows exactly what to do when the warning sounds.

This story encapsulates several of the lessons we must learn in adapting to climate change. Part of the answer will come through engineering projects, part through warning systems and preparedness. The rest, perhaps the hardest part, will require changes in many of the ways that we currently make our living.

Engineering Projects

One of the most important roles that engineering projects will play in helping to adapt to climate change will be to provide protection against coastal flooding (through rising sea level and increasing storm surges) and inland flooding (because of more intense rainfall). The United Kingdom is now in the process of implementing a nationwide strategy for flood and coastal defenses, based on a detailed Foresight report commissioned by one of us (David King) in 2002. We learned many useful lessons from this exercise.

First, there is the importance of accurate—and regionally detailed—model predictions. The Hadley Centre models that we were using had a resolution of 160 by 160 miles. In other words, the results came out as uniform squares with sides of 160 miles. On a global scale, that tells you a lot. But if you start to use boxes that size to try and draw conclusions about an area the size of an individual country, you end up with almost no regional detail. What we really need for good adaptation strategies for individual cities, counties, and regions are global models that can make accurate climate predictions on a 40-by-40 mile scale, or even lower. That sort of resolution is in the pipeline but will require the world's largest computers to implement. It should be an urgent priority. Note that what's important is not so much a detailed prediction of where more rain will fall. Instead, the models will identify the areas that are most vulnerable to likely changes and extreme events.

The next step is to design the barrier, levee, or drainage systems to cope with the changes that the models predict. This is extremely important. Hurricane Katrina was certainly strong, but it wouldn't have had the same deadly impact had the levees been higher and stronger. New Orleans's defenses were designed to cope with an event that would be expected to occur once in every two hundred years, which wasn't nearly enough even before climate change started raising the baseline.

This illustrates an important aspect of the rhetoric about climate change. In spite of what you might hear from some of the more fervid doomsayers, the reason that hurricane damage has risen in the past few decades is not, in fact, more intense hurricanes. Even though hurricanes do seem to have been getting stronger, a much bigger factor to date has been the rush to the coast. There has been a massive increase in the numbers of people living on the world's coastlines and a minimal amount of attention paid to protecting them. The increasing storm intensity and rising sea level that we can anticipate in the future will only make this situation worse.

Once the projects are designed, it's necessary to test whether they are likely to be effective. The United Kingdom's Foresight project designed a computer program called Floodranger, which was based on the computer game SimCity. The program allowed designers to "build" their project, run the climate forward, and see how the defenses coped. The value of this approach became clear when the model produced some surprising results. For instance, it suggested that the coasts weren't the only problem. Floodranger showed that inland cities could also suffer if flash floods from more intense rainfall overwhelmed their drainage systems. This is exactly what happened in the British floods of 2007, during which more than twenty people died and tens of thousands were washed out of their homes and businesses.

Finally, there need to be funds to pay for the adaptation. The British government is now putting £500 million (more than $1 billion) per year into an urgent program to upgrade coastal and drainage defenses, and this will rise to £700 million (roughly $1.4 billion) per year in 2011.

Similar national programs will need to be put in place for vulnerable regions around the world. The United Kingdom is now working with China to transfer the lessons learned to Shanghai and the Yangtze basin. Many of the most vulnerable countries are also the ones least able to pay for national programs on this scale; we will talk more in chapter 10 about how these can be funded.

Projects like these should also be constantly revisited as the models progress. In the case of the Foresight project, this will happen every five to ten years.

Meanwhile, the message is beginning to filter through to many parts of the world that any major new civil engineering project will need to take climate change into account. For instance, the eight-mile Confederation Bridge in Canada, which connects Prince Edward Island to the mainland, was built three feet higher than is currently needed, to take into account the possible rise in sea level over its hundred-year life span.[8]

Similarly, Boston's Deer Island sewage treatment plant has been built on a high enough elevation to protect its output pipes from being inundated by rising seawater. Up on the Tibetan Plateau, the Qinghai–Tibet railway is vulnerable to melting permafrost. More than three hundred miles of the railway rest on permafrost that is within a few degrees of thawing, and engineers have incorporated into the design a combination of insulation and cooling systems to ensure that the trains will not tip the permafrost over its melting threshold.

The Konkan Railway in western India, the Copenhagen Metro in Denmark, and the Thames Barrier in London are all currently undergoing assessment for incorporating climate change into their designs.

Sound the Alarm and Know the Drill

This approach will be necessary for the sort of sudden, dramatic events that can cause considerable loss of life and damage to property if they take us unawares. Tragic examples that we have already experienced include the 2003 heat wave in Europe, the deaths from Hurricane Katrina in New Orleans in 2005, and the wildfires that scorched Greece in 2007. A combination of effective warnings and preparedness for these events could have saved thousands of lives.

When it comes to early-warning systems for natural disasters, one of the best examples is the one used in Bangladesh for an impending typhoon. The World Meteorological Organization issues a warning that a severe storm is on the way. Men on bicycles with whistles transfer this information to the populace. The local people have been drilled to move immediately to higher ground or to high-rise shelters. Though typhoons in the region seem, if anything, to have been growing more severe, death rates have plummeted.

This illustrates the two important aspects of any effective warning system. First, the alarm has to be official, provided by one overseeing and reliable body that will vet any potential threat. Above all, we can't afford false alarms, which allow people to become complacent.

The WMO has already agreed to work with the United Nations to extend its warning system beyond hurricanes and typhoons to all physical natural disasters. We propose that it would be natural to include extreme weather in this program. The WMO would be the most effective body to oversee a global warning system for events linked to climate change.

Second, everybody has to be clear about exactly what to do. In central Europe in 2003, many people—including hospital staff—were on holiday when the heat wave struck. There was nobody to issue simple information to vulnerable older people, such as instructions to drink enough water to avoid dehydration or to go out only during the coolest parts of the day.

While we're talking about adapting to heat waves, it's worth noting where the danger actually lies. We humans are already amazingly adaptable to the temperatures of our local environment. One study discovered that the ideal temperature range in northern Finland to minimize heat-related deaths was 58°F to 63°F, while the temperature that the people of Athens coped best with was a much hotter 73°F to 78°F. The message here is that we survive best at the temperatures to which we're most accustomed.[9]

Much of how we adapt has to do with habits, such as what time of day we typically choose to do outdoor work or whether we drink water before leaving the house. The rest depends on how our infrastructure is set up, such as the sorts of buildings we live in and the means of transportation we use outside. Habits are easier to change than buildings, but neither of them typically changes very fast.

Thus the problem with a heat wave (or, indeed, a cold spell) is that it takes the temperature above (or below) our comfort zone more quickly than we can reasonably respond. That's why so many people died in Europe in 2003. The challenge now is to adapt our habits and our buildings beforehand so that we can be ready when the heat waves come.

Gradual Loss of Ecosystem Services

In some ways, this is the hardest adaptation to quantify. In this category go different gradual changes in our capacity to extract a living from the land. We have already described many of these above, such as changes in water availability, temperature, or a shift in the pattern of pests or illnesses. What's needed is a big improvement in the models to identify the regions that are most vulnerable to these changes and come up with strategies to cope. These might include changing crops to varieties that are most resistant to drought, and perhaps using biotechnology to help develop new, more resistant strains; changing planting dates and irrigation methods; harvesting and storing rainwater more effectively; using controlled burning to reduce fire risks; and planting mangroves to provide natural seawalls against the rising waves and to reduce loss of soil from the coast.

Approaches like these will need to be integrated into a national strategy for each country in the world. It's not enough to build up the seawalls and hope for the best. Governments will need to know exactly what climate change will mean for their countries, and how they will need to respond.

Some climate changes might overwhelm our capacity to adapt. If your land is in one place and your experience lies with the crops you have traditionally grown there, it may not be easy to switch to something different. Social, economic, political, and cultural barriers all stand between the world we have now and the one that we will soon have as climate takes its toll. And in some cases, such as for people living on the lowest-lying atolls threatened by rising sea levels, the only recourse might be to move.

It is already too late to stop the changes that we describe in this chapter. These adaptations, and more, will be necessary no matter what we do now to reduce the impact of future warming. Part 3 will look more closely at how the adaptations will be paid for, and how they can be politically encouraged. But for now, it's worth noting that no part of the world can be treated in isolation. The climate change that is already in the pipeline will hit hardest those countries that are least equipped to deal with it—and that have been least responsible for releasing the emissions that caused the problem in the first place. Because of this, we believe that developed countries have a moral responsibility to help the developing world cope with the coming damage. But there's an even stronger reason why we need to pay serious, united, international attention to adaptation. From the security of our food and water supplies to the economic implications of massive environmental migration, when global warming affects anybody, those effects can quickly lead to trouble elsewhere.

5

CLIMATE WILD CARDS

If the world will now need to adapt to change regardless of what we do, can't we just keep on burning fossil fuels and simply prepare ourselves for the consequences? That would work only if we were completely confident of being able to take anything Earth's climate can throw at us. The trouble is that there are many ways in which it can take us by surprise.

The opening scenes of the Hollywood climate disaster movie *The Day After Tomorrow* show the hero working in Antarctica. He is camping on a massive floating shelf of ice, called Larsen B, drilling ice cores through its surface. The first sign of impending doom comes when the ice shelf cracks, right across the research site. In trying to save his precious ice cores, our hero only barely survives.

This scene is based, loosely, on a true story. Larsen B is a real ice shelf—or at least it was. It was more than one thousand square miles in area, and seven hundred feet thick. In February 2002, over a period of just a few weeks, Larsen B unexpectedly shattered. By the beginning of March, five hundred million tons of ice, the equivalent to an area bigger than Rhode Island, had vanished.

Though scientists had already noticed that this part of the Antarctic was getting steadily warmer, the Larsen B breakup happened more quickly than anyone had expected. In fact, it happened more quickly than even the Hollywood scriptwriters could handle. One scientist remarked after seeing the film that in real life the cracks would have been happening all around, and the hero would have had to run for his life. This was perhaps the only example of a Hollywood disaster film that underplayed reality.

However, many of the movie's subsequent events managed to be both exciting and scientifically ludicrous. For instance, mighty hurricanes formed over land, whereas in fact they can only be born over water. And a supposedly climate-driven tsunami swamped the streets of New York, although such tidal waves cannot be caused by rising temperatures.

As global warming awareness has taken hold, disaster scenarios like these have become increasingly popular. There have been many books, articles, and films about impending climate disasters, some of which are just as lurid as *The Day After Tomorrow*. The question is, are they true?

There are certainly some wild cards in the climate, and there is every chance that if we don't curb emissions, the result will be even worse than a mere multiplying of the effects that we described in chapter 3. In this chapter we will separate the myths from reality, starting with the least likely and working our way upward.

Shutdown of the Ocean's Circulation

This is perhaps the scenario most beloved of disaster-mongers, and is also the one at the heart of *The Day After Tomorrow*. Despite their appearances, the world's oceans are not chaotic individual masses of water. Beneath their surfaces lies a supremely organized conveyor belt of currents that transports unimaginable quantities of water around the world. There is at least one region where this conveyor belt can be switched on and off: a few relatively small patches of water located between Greenland, Iceland, and the north coast of Norway.

In that part of the world, seawater moving north at the

surface experiences two important changes. First, it becomes very cold. Second, it becomes much saltier, because the top part of the ocean begins to freeze into "fresh" sea ice, leaving much of its salt in the water beneath.

Cold, salty water is much heavier than normal seawater, and so begins to sink. As it sinks, it pulls surface water in to replace it, just as when you take out the plug in a bath, surface water races toward the end of the bath to replace what's disappeared down the drain. In this case, the descending arm of the conveyor belt drags balmy tropical waters—the famous Gulf Stream—up north, where they lap the shores of several northern European countries and help to explain why these are much warmer than parts of North America at the same latitude.[1]

In the scenario described in *The Day After Tomorrow*, a shutdown of the conveyor belt stopped the Gulf Stream from heading north, and hence brought a spectacular ice age to northern Europe and North America. However, although there were early fears that this might indeed take place, scientists now believe that switching off the conveyor belt would only lower European temperatures by a few degrees, and would probably just barely cancel out the global warming there.

This sounds as if it would be a good thing. But the sinking and dragging northward of surface water sets in motion an entire network of oceanwide currents that stretch down the Atlantic, around the southern coast of Africa, through the Indian and Southern oceans, and around the Pacific. This gigantic conveyor belt carries not just water but salt and warmth right around the globe. Switch it off, and you have the potential to change climate worldwide.

For instance, similar changes in ocean circulation in the past seem to have significantly changed the pattern of tropical rainfall, including the monsoons that ultimately provide much of Asia with its food.[2] Europe and North America would also not escape scot-free. Changing the ocean currents in this way would redistribute seawater, making the North Atlantic rise by up to three feet, in addition to the rise that had already come from melting ice.[3] It's also possible that the shifting currents would change the location of the food supply for schools of fish and cause lucrative Atlantic fisheries to collapse.

The best way to switch off the conveyor belt would be to melt large amounts of ice right on top of the main sinking point. That would leave a freshwater "lens" that could act like a plug, stopping the water from getting dense enough to sink in the first place. That's exactly what's already happening in the Arctic. Both the floating sea ice and the land-based Greenland Ice Sheet are delivering freshwater exactly where it could hurt.

The good news is that, as yet, they're not delivering nearly enough. Ocean modelers have compared the output from eleven different models and have concluded that the amount of freshwater needed to shut down the North Atlantic conveyor belt would be about the same as that now coming from the Amazon.[4] At the moment, the Arctic has hardly reached one-quarter of the model's lowermost threshold. And of course the sea ice won't be around for much longer to keep contributing.

The IPCC report states that it is extremely unlikely that the conveyor belt will shut down in the twenty-first century, whatever we choose to do about climate change. However, if Greenland starts to melt more quickly (see below), there is still a chance that the conveyor belt could be vulnerable in the longer term. In the meantime, there are already signs that the system may be weakening, and models predict that if we do nothing to halt carbon dioxide emissions, the conveyor belt could be 50 percent weaker by the end of the century. The danger is that this might bring a similar pattern of effects as for a total shutdown, albeit of a lesser magnitude.[5]

Likelihood: The collapse of the ocean conveyor belt is very unlikely over the next century, and improbable even in the following
hundred years. There is also virtually no chance that a shutdown
would cause a European ice age. However, if we don't constrain
climate change, weakening of the circulation by the end of this
century could still reduce tropical rainfall, including the Asian
monsoons, as well as raise sea level in the North Atlantic.

Massive, Abrupt Sea-Level Rise

The sea is already rising, as melting glaciers add their load to
oceans that are expanding as they warm. But what if the world's
ice sheets tipped over a threshold and suddenly, irreversibly, slid
into the sea? There are three major ice sheets to worry about,
two in Antarctica and one in Greenland. So far, these three together contribute only about one-tenth of the current global
sea-level rise of one-tenth of an inch per year. But the signs are
that the melt could be accelerating, and there is a frighteningly
large amount of ice at stake.[6]

Antarctica

Antarctica's two great ice sheets are linked at the center like
slightly misshapen butterfly's wings. The larger of these, the East
Antarctic Ice Sheet, contains enough fresh water to raise sea
levels by more than one thousand feet. However, nobody thinks
it will melt in the near future. The eastern sheet is old, cold, and
set in its ways, and recent satellite data have even shown that it
is putting on weight rather than losing it.[7] (That's not a contra-
diction of the global warming story, by the way. The models predict that, in one of the many counterintuitive consequences of
warming, as the seas warm and sea ice melts around Antarctica,

more water will be sucked up into the air, and this will then fall as extra snow in the frozen interior. Unfortunately, the amount of extra snow captured is still only enough to slow sea-level rise by about one-hundredth of an inch or less per year.)

However, the smaller, western sheet is more vulnerable, because most of its glaciers sit on a thin film of water rather than being frozen to the rock like the eastern half. If the West Antarctic Ice Sheet slid into the sea, it would still raise global sea levels by a horrifying sixteen feet.

Clues as to how this might take place come from the doomed Larsen B ice shelf. The collapse of Larsen B isn't necessarily a sign that global warming has already come for the rest of Antarctica. Its home was the Antarctic Peninsula, a finger of land that sticks out toward South America and that is so northerly and relatively balmy that old Antarctic hands refer to it as the "banana belt."

Though the peninsula has certainly been getting even warmer lately, with temperatures rising by around 0.9°F a decade since the 1950s, the rest of the continent has remained much more stable. And though recent evidence suggests that Larsen B had previously remained intact for at least ten thousand years,[8] it is still possible (though increasingly unlikely) that the warming there is merely part of a natural fluctuation. Also, the collapse of Larsen B made not one iota of difference to sea level. Floating ice doesn't raise or lower the water level as it melts. You can see this for yourself by putting an ice cube in a glass of water. As the cube melts, the water stays at exactly the same level in the glass.

But, disturbingly, behind Larsen B were a set of flowing glaciers that the ice shelf had previously held back. With the shelf gone, these glaciers have now started to accelerate.[9] When glacier ice flows from land into the sea, it adds water that wasn't there before. In other words, it raises sea level.

Those glaciers are relatively small by Antarctic standards. But there are much, much larger ones draining ice from the mighty West Antarctic Ice Sheet into the sea, and all of these are held back by two vast floating ice shelves that dwarf Larsen B. If these shelves were to shatter, it's conceivable that most of the West Antarctic Ice Sheet could then slide into the sea.

Fortunately, these gigantic ice shelves are much less vulnerable than Larsen B was, since they are at least five times thicker. Few researchers think they will break up soon, even if climate change is allowed to go unchecked. But they are still being very closely monitored, just in case.

The glacier/ice shelf systems make up roughly two-thirds of the West Antarctic Ice Sheet. Until recently, the remaining third was relatively unstudied, largely because the weather there is awful even by Antarctic standards. However, this region, called Pine Island/Thwaites, contains three of the continent's largest glaciers, which drain ice from the high interior out into the sea. Because the front ends of these glaciers are exposed directly to the sea, without the benefit of shelves of ice to buffer them, they have been called the "soft underbelly" of Antarctica. And there are worrying signs that these glaciers are speeding up and draining more ice.[10]

Likelihood: The massive East Antarctic Ice Sheet is virtually certain to stay intact, and might continue to soak up some sea-level rise through extra snowfall. Most of the West Antarctic Ice Sheet is unlikely to collapse sufficiently to provide an abrupt dramatic rise in sea level this century, though this conclusion depends on aspects of its internal dynamics that are as yet unclear. The Pine Island/Thwaites glaciers could be more vulnerable. How likely they are to collapse remains uncertain, and is now a subject of urgent study. If they collapsed, taking with them the ice they drain from the interior, global sea level could be raised by about five feet, drowning much of the world's coastlines.

Greenland

Melting the entire Greenland Ice Sheet would increase sea levels worldwide by twenty-three feet. Because the Arctic is warming much faster, it is probably more vulnerable than its Antarctic cousins.

But is it melting? In the past few decades the picture has been anything but clear, with different measures giving different answers. In some places a glacier was receding, while in others one was surging forward. But of late all the signs have begun to point in the same direction. Greenland is indeed beginning to melt.

Though satellite measurements of Greenland's interior suggest that snow has recently been building up there, the edges of the ice sheet are also getting thinner.[11] Greenland's mighty outlet glaciers seem to have been speeding up, too, more than doubling its annual loss of ice over the past decade.[12]

The good news is that this process has not yet doomed the Greenland Ice Sheet to extinction. In 2007 modelers calculated that it would take a global average warming of 5.4°F to push Greenland over the edge,[13] a threshold that we probably still have time to avoid.

But there is another important factor that the models didn't include. In 2002 several researchers noticed that something alarming seemed to be happening in one region during the summertime. Just when the summer sun began melting parts of the surface of west-central Greenland, the glacier there began to slip more quickly,[14] even though Greenland's glaciers are many hundreds of feet thick. And because they slip on their bellies, something that happens on the surface shouldn't make any difference. It turned out, though, that the melting was producing giant lakes on the surface, whose water then poured down through cracks and crevasses, creating inner waterfalls that flow to the land below and lubricate the ice's sliding edge. That's worrying because, in principle, it means that the entire Greenland Ice Sheet could be more vulnerable than we think to the steadily warming air.

The models don't incorporate this mechanism because they can't. Nobody yet knows how widespread the effect might be, or even where the cracks are. What's more, even before it reaches the point of no return, Greenland can still raise sea level by amounts that we humans would find catastrophic. The ice sheet didn't disappear during the last warm period, around 130,000 years ago, when temperatures in the north were a few degrees higher than they are today. And yet the latest analyses suggest melting Greenland ice increased the sea level then by somewhere between six and ten feet. That would be more than enough to cause serious trouble in today's world.

Likelihood: The Greenland Ice Sheet is already melting. Unconstrained climate change, with a global average temperature rise of 5.4°F, could doom the ice sheet to eventual disappearance within the next few centuries, which would condemn future generations to a sea-level rise of twenty-three feet (though even if it tipped over this threshold, the entire ice sheet would still take many more centuries to melt entirely). However, as with the West Antarctic Ice Sheet, this assessment of the vulnerability of Greenland depends on aspects of its internal dynamics—for instance, whether the glaciers will begin to slide more quickly as surface meltwater falls through cracks—that are as yet uncertain. If these mechanisms cause Greenland to melt more quickly than we expect, sea level could rise by many feet over the next century, which would pose grave danger for our civilization.

Greenland is one of the most convincing reasons we have for the urgent need to curb climate change.

Melting Permafrost

In the frozen wastes of the Arctic lies a climate wild card that makes even the soberest scientists afraid. From Alaska and northern Canada across the northernmost parts of Europe and Siberia lie millions of square miles of frozen earth. In some places the land is forested with black spruce; in others there are sedges, mosses, frozen lakes, and marshes. But around the fringes, and sometimes in the heart of this freezing land, the thaw has already begun.

Spruces in northern Canada have begun to lurch drunkenly as the ground falls beneath their feet. Giant holes have opened up in Alaska as the ice that once held the soil together turns to water and trickles away. Siberia's lakes have begun to awaken from their frozen sleep. And they are all, inexorably, putting out carbon.

This is the Arctic permafrost, so called because it contains a layer of soil that stays frozen through summer and winter. This soil acts like a freezer, trapping carbon in the form of leaves, roots, dead mosses—anything that was once alive. But the power to the freezer has been turned off, and the carbon inside is beginning to rot.[15]

The danger is that the warming caused by our own human emissions of carbon dioxide is triggering a thaw of the north, and that that in turn will emit yet more carbon of its own in a vicious cycle that accelerates the warming. This is another example of a positive feedback, positive not because it's good but because it exacerbates the original effect.

In this case, what's frightening is the sheer scale of the potential problem. Some researchers have estimated that there could be a full nine hundred gigatons of carbon waiting to be released from the rotting Arctic soils, which is more than we currently have in the entire atmosphere. If even a small percentage of that escaped, it could double or triple the effect of our human emissions. And then all climate bets would be off.

Also frightening is that when it comes to thawing permafrost we have very few facts to go on. Until recently, research efforts have been too patchy and sporadic to put the whole picture together. Now Arctic scientists are scrambling to understand exactly what will happen.

We do know that the Arctic is already warming up. Temperature measurements show that not a single part of the region is cooling down and most parts are seriously heating up, in some places at more than double the global average rate.[16] We also know that the permafrost is already thawing in the fringe regions where it is at its thinnest and patchiest. There are also worrying signs that parts of the much colder interior may have begun to release their carbon, as bubbles of methane emerge from lakes in Arctic Russia.[17]

What's more, the form in which the carbon comes out can make a big difference. In the muddy marshland of northern Sweden, the thaw has actually decreased the overall amount of carbon emerging, because the water has blocked off the soil from the air, which in turn slows down the rotting process. But much of the carbon that does come out is now in the form of methane—marsh gas—which packs a much greater punch in the global warming stakes than carbon dioxide. The upshot: Though the overall amount of carbon released has dropped by 13 percent, the warming effect has gone up nearly 50 percent.[18]

Likelihood: uncertain but very worrying. Nobody yet knows what will happen to permafrost in the future. The first tentative attempt to model the effect suggested that if emissions continue unabated, by the end of the century 90 percent of the permafrost will be gone. This figure is probably an overestimation, and better models will be available in the next five years or so. But if even a small percentage of the available carbon comes back out of the ground, that would effectively double our current human emissions at a stroke. And the resulting climate disaster doesn't bear thinking about. Though it has more or less escaped the attention of the writers of disaster scenarios, permafrost is the most frightening wild card in the deck.

Something We Don't Know About Yet

It might sound facetious, but the biggest dangers could come from a disaster scenario that nobody has yet noticed. The climate records written in ice and rocks suggest that our atmosphere is supremely temperamental, capable of tipping over a dramatic edge into a new state in a very short amount of time. For instance, at the end of the last ice age the sudden melting of vast amounts of ice shut off the ocean circulation, which caused an abrupt refreezing. Though it's still not clear exactly how global the event was, Greenland's temperature dropped by 27°F, and northern Europe's by 9°F. The freeze happened over several centuries, and at first scientists studying ice cores thought the thaw had taken equally long. But as new, more detailed ice cores were drilled, researchers discovered to their astonishment that the shift back to warm temperatures at the end of this deep freeze probably happened within a decade. (As one researcher put it, dryly, the climate shift took place within a "congressional lifetime.")

What's more, similar abrupt lurches from cold to warm and back to cold again seem to have happened throughout the ice age itself, and in spite of intense research nobody is yet sure exactly why. And then there's the time, around fifty-five million years ago, when some unknown event apparently triggered the release of massive amounts of carbon, which in turn raised global temperatures abruptly by some 14°F.

In this context, it's alarming to think that, thanks to our human burning of fossil fuels, carbon dioxide levels are higher today than they have been for at least 650,000 years and probably much longer. Renowned climate scientist Wally Broecker from Lamont Doherty Earth Observatory has likened the climate to an angry beast that we are now poking with a stick. If we keep on poking, we don't know what will happen next. But we do know that it could happen swiftly enough to take us completely by surprise.

Even without these disaster scenarios, the IPCC report shows that we have ample reasons to fear unabated carbon dioxide emissions. The changes that we described in chapter 4, and which are now inevitable, are just a hint of what we would have in store if climate change were allowed to continue unabated for the next century or more.

This alone should provide us with a serious incentive to stop our greenhouse emissions before things get worse. But the wild cards that we have described in this chapter put into even starker perspective the arguments for kicking our carbon habit. Take sea level. The IPCC report suggests that if carbon dioxide emissions continue as foreseen, by the end of the century sea level will have risen by between one and one and a half feet. Especially at the higher end of the range, that would bring the danger of flooding to millions more people.

It's also worth remembering that the IPCC is inherently a conservative body. Some researchers have since suggested that, if anything, this prediction is likely to be an underestimate. In particular, it doesn't include any of the mechanisms by which the ice sheets could crumble. If all the possible ice mechanisms combine to conspire against us—if, for instance, meltwater at the surface does indeed flood down to the base;

if that sends the ice sliding ever faster on Greenland and West Antarctica; if the buttressing ice shelves do start to crack; and if warm sea begins to lap at the vulnerable mouths of the Pine Island glaciers—then there is a serious chance that by the end of the century the sea will have risen by a matter not of inches but of feet.[19] That's not a chance we should want to take.

Another potential danger lies in our reliance on the environment's capacity to soak up the carbon dioxide that we emit. To date, the oceans and the ecosystems on land have soaked up about half of the emissions that we humans have put out. If it weren't for this particular ecosystem service, we would now be experiencing much more severe effects from climate change. But at some point in the future, the natural world might become saturated and lose its capacity to soak up emissions. When and whether that will happen in the oceans is a complicated question. The ability of the oceans to take up carbon dioxide depends on their chemistry (how warm and acidic they become), their physics (whether ocean circulation patterns change), and their biology (what happens to the organisms that trap carbon in their bodies and are then buried in the seafloor). No model has yet managed to give a convincing answer to how these different effects will combine.

On land, the picture is a little clearer, but also bleaker. Even without considering the effect of thawing permafrost, the IPCC report suggests that the capacity of land-based ecosystems to soak up carbon dioxide will saturate around the year 2030. If we don't restrict emissions, by 2070 plants and soils may even become a net source of carbon dioxide—which would add to our human contribution of greenhouse gases and accelerate the warming.[20]

If we continue to burn fossil fuels with abandon, the more distant future will be much worse than anything we've described so far. There's enough fossil carbon in oil, gas, and—especially— coal to keep the world going for centuries. Long before it's exhausted, we will have passed the threshold for all the eventualities described in this chapter, and more. In the end the atmosphere could contain thousands of parts per million of carbon dioxide, a level not seen since the time hundreds of millions of years ago when hyperhurricanes lashed Earth's surface with acid rain.

There are plenty of excellent, well-researched books that have looked into this disturbing crystal ball.[21] Here, we prefer to assume that mankind will not be foolish enough to let climatic disaster happen. Our generation is the last to have the chance of averting the worst of these scenarios. All we need is the right combination of new technologies and economic, political, and social will. The next two sections of the book will look at each of these in turn.

PART II

TECHNOLOGICAL SOLUTIONS

Now that we know the scale of the problem, the next question is what to do about it. Applying existing technologies and developing new ones will be a large part of the answer—though that's not all it will take. In this section of the book we explain what low-carbon technologies and strategies are already available and what else is on the horizon. The tools we describe will need to be used on every scale, from the personal to that of cities to that of countries. They will involve many different sectors of the economy, different greenhouse gases, and different ways to produce and use energy. We'll explain how these technologies are likely to affect your life, and at the end of the book we'll outline specific ways in which your individual choices about employing these approaches will make a difference. First, though, we need to establish a target for the ultimate levels of greenhouse gas at which we should be aiming.

6

WHAT SHOULD WE AIM FOR?

To tackle climate change we will have to reduce our emissions of greenhouse gases. There is no other way. But what figure should we aim for? This is the first and most important question to answer. Until we have a specific target, we will have no idea what technological or political solutions will enable us to achieve it.

Many previous attempts to decide how far we should allow our climate to go began with the idea that we should stop climate change before it becomes "dangerous." However, it's too late for that. As we explained in chapters 3 and 4, climate change has set the stage for many dangerous events that either have already taken place or are now inevitable.

A better question is how much climate change we can afford before things become truly catastrophic. Though the answer is full of scientific uncertainties, a broad consensus is beginning to emerge.

How Hot Is Too Hot?

The global average temperature has already risen by some 1.5°F since the nineteenth century. If we switched off every fossil-fuel power station, grounded every plane, car, and train, and went to huddle around wood fires, there would still be another 1°F of warming to come as the atmosphere catches up with the greenhouse gases we have already put into the air. Thus the minimum possible temperature rise for the world, compared to preindustrial times, is approximately 2.5°F. There is nothing we can do to stop at least that much. Below, we outline some of the changes predicted by the IPCC report for warming of 3.5°F (2°C) and upward.[1]

Vulnerabilities at Warming of up to 3.5°F (2°C)

❑ Higher global crop yield than today, but this masks an inequality. In some middle- and high-latitude countries crop yields go up, but in the tropics they are already falling. Between 10 and 30 million more people will be at risk of hunger.

❑ Increase in human health problems from heat waves, malnutrition, floods, droughts, and spread of infectious diseases.

❑ Less water availability and more droughts in the middle latitudes and semi-arid tropics. Between 0.4 and 1.7 billion people will be suffering from increased water scarcity.

❑ Environmentally driven migration with the potential to exacerbate conflicts over scarce resources and cultural invasions.

❑ More intense individual rainfall events, causing potential for flooding even in regions that are otherwise suffering from more severe droughts.

❑ Increase in intensity of hurricanes.

❑ Increased heat waves in continental areas, as well as more droughts and fires in midlatitude continental areas as the storm tracks move poleward.

Additional Vulnerabilities at Warming of 3.5–5.4°F (2–3°C)

All of the above, plus:

- ❑ Up to 3 million more people at risk of flooding.

- ❑ Up to a further 10 million people at risk of hunger.

- ❑ Most of the world's coral reefs bleached.

- ❑ Commitment to widespread de-glaciation of the Greenland and perhaps the West Antarctic ice sheets, with potential sea-level rise of more than ten feet.

- ❑ Considerable weakening of the ocean conveyor belt, with potential for significantly reducing monsoon rains.

- ❑ Further increase in intensity of hurricanes, enough to exceed infrastructure design criteria, cause significant economic loss, and threaten large numbers of lives.

- ❑ More flooding in North America and Europe as winter rainfall increases and less water is stored as snow.

- ❑ Rapid increase in frequency of serious heat waves, causing many deaths as well as crop failures, loss of forest, and fires.

- ❑ Extreme drought in increasingly larger areas.

- ❑ Serious threat of inundation in low-lying coastal areas and small islands.

- ❑ Accelerated shrinking and eventual loss of tropical mountain glaciers.

- ❑ Considerably more environmentally driven migration.

- ❑ Between 20 and 30 percent of all species on Earth at increasingly high risk of extinction.

Additional Vulnerabilities at Warming of 5.4–7.2°F (3–4°C)

All of the above, plus:

❏ Increasing likelihood of near-total melting of Greenland and West Antarctic ice sheets, leading to eventual sea-level rise of forty feet over the coming centuries.

❏ Switch from terrestrial ecosystems soaking up carbon to being a net source of carbon, accelerating the warming rate.

❏ Major species extinctions around the world.

❏ Widespread death of coral reefs.

❏ Falls in food yields in some parts of the higher latitudes. Global food production begins to fall.

Additional Vulnerabilities at Warming of 7.2–9°F (4–5°C)

All of the above, plus:

❏ Falls in food yields even at the most favorable locations in higher latitudes. Plummeting global production.

❏ Increasing risk of serious, abrupt changes to the climate system, including shutdown of the ocean circulation and major release of carbon from thawing permafrost.

❏ Up to one-fifth of the world's population affected by flooding.

❏ Up to 120 million more people at risk of hunger.

❏ Between 1.1 and 3.2 billion people suffering from increased water scarcity.

❏ Loss of almost all high-latitude forest and very large tracts of the Amazon rain forest.

❏ Ecosystems covering 40 percent of the world's land area subject to major changes.

❏ Very substantial increases in human deaths from malnutrition, infectious diseases, heat waves, floods, and droughts.

Additional Vulnerabilities at Warming of 9–11°F (5–6°C)

❏ The IPCC didn't want to go there, nor do we.

Pick a Number

Looking at the above lists, none of the outcomes is desirable, and all become increasingly less so as the temperature scale rises. All things being equal, if forced to choose among them, to keep the "danger" as low as possible we would pick the lowest possible rise—in other words, set a temperature limit of 3.5°F (2°C). In addition to the 2.5°F that's already inevitable, that would allow us only 1°F of leeway in which to kick our carbon habit.

Even before the advanced model predictions in the IPCC report were available, researchers had been trying to determine how much warming humans can bear. Perhaps surprisingly, and through quite different reasoning, many of them have hit on exactly the same figure: 3.5°F.

One early strategy was to avoid "dangerous" climate change by keeping our warming to within the upper limits of natural variability during the relatively stable ten thousand years since the end of the last ice age. Pioneering economist of climate change William Nordhaus from Yale University suggested in 1979 that this limit might be on the order of 3.5°F, though he admitted there was a great level of uncertainty in the data.[2]

In 1995 the German Advisory Council on Global Change (WBGU), an independent science council set up by the German government in the run-up to the Rio Earth Summit, published a report suggesting that the highest global mean temperature in the past several hundred thousand years had been about 2.6°F above preindustrial levels. They added an extra 0.9°F on the principle that the Industrial Revolution has at least brought us more capacity to adapt to a changing world, and proposed the same maximum temperature limit of 3.5°F.

This isn't necessarily the best way of deciding on a "safe" temperature rise. For one thing, during the last warm gap between ice ages, which occurred about 130,000 years ago, the global mean temperature was only a whisker higher than it is today. But thanks to slight wobbles in Earth's orbit, the heat was distributed differently, the Arctic grew much warmer, ice melted, and sea levels were up to ten feet higher than they are now. Looking at the past can be a useful guide to the future, but it's still not a mirror image. Moreover, temperature rises that occurred before humans existed—and long before there were more than six billion of us living in our fixed cities and depending on our industrial lifestyles—have little to say about what we might find "dangerous" today. However, the more we discover about the dangers of our interference with the climate system, the better that almost accidental figure of 3.5°F looks. The above lists show that the most serious effects begin to kick in somewhere between 3.5°F and 5.4°F and become increasingly dangerous as the numbers go up.

In 2003 the WBGU produced another report,[3] which set out the specific dangers likely to be faced for different temperature windows. They concluded that, based on dangers such as floods, famine, heat waves, droughts, and extinctions, temperatures should not be allowed to rise beyond 3.5°F. The report added: "This climate window should be agreed as a global objective . . . The European Union should seek to adopt a leading role on this matter."

This conclusion was not limited to green Germany. In 2005 the International Climate Change Taskforce, a group of scientists and politicians from around the world, published a report.[4] The task force was cochaired by British member of Parliament Stephen Byers and American Republican senator Olympia Snowe, and, as well as the usual concerned faces from the industrialized world, also included members from China, India, Malaysia, and Australia. Their conclusion? To avoid the worst effects of agricultural losses, water shortages, costs to health, damage to coral reefs and terrestrial ecosystems, and the increasing risk of passing a dangerous climate tipping point, "We propose a long-term objective of preventing average global surface temperature from rising by more than 2°C (3.5°F) above its preindustrial level."

That same year, the European Union confirmed the adoption of this figure as its official policy. The entire continent is now committed to attempting to keep global temperatures from rising more than 3.5°F above preindustrial values, and to persuading the rest of the world to follow suit.[5]

Now for the Really Bad News

The very serious problem that few people appear yet to have noticed is that it's now almost certainly impossible to restrict warming to 3.5°F. If we had started two decades ago we would have had a good chance, but in the present climate that target looks increasingly out of range.

To guarantee staying below a given temperature we need to know exactly how sensitive the climate is to rises in greenhouse gases—how much temperature "bang" you get for every greenhouse "buck." Our estimates for this sensitivity are getting better all the time, but they're still uncertain. If the sensitivity is at the low end of the range, we still have a slight chance of staying below 3.5°F. But if it's at the high end, even 5.4°F may already be beyond us.

Whether the climate is thick-skinned or high-strung depends on the various feedbacks that act with carbon dioxide to boost its effect. For instance, as we described in chapters 1 and 2, warmer air soaks up more water vapor, which is a potent greenhouse gas in its own right, and thus causes even more warming. Also, melting shiny white sea ice and replacing it with dark seawater enables the world to soak up more sunlight, causing more warming. The strength of all of these feedbacks added to the power of carbon dioxide and the rest of the greenhouse family together determine the climate's sensitivity.

Because they all treat the feedbacks in slightly different ways, climate models come up with a variety of numbers for the sensitivity. Putting these together gives a range of probable temperatures for any given amount of greenhouse gas in the air. Forgive us for presenting yet another blizzard of figures, but they are vital to the story:[6]

❑ For 450 ppm CO_2eq, the temperature rise will probably be between 3.5 and 6.5°F, with the likeliest value about 4.5°F.

❑ For 550 ppm CO_2eq, the rise will be between 5.5 and 9°F, with the likeliest value about 6°F.

❑ For 650 ppm CO_2eq, the rise will be between 6 and 11°F, with the likeliest value about 7.5°F.

These numbers should ring very loud alarm bells. The concentration of greenhouse gases in today's atmosphere is currently about 430 ppm CO_2eq. This means that 450 ppm is the lowest we can hope to achieve. And yet look at the range. The *likeliest* value for the temperature rise even if we manage to stay at 450 ppm is already 4.5°F. If we're very lucky, and the sensitivity is at the lowest part of its range, we might stay below that.

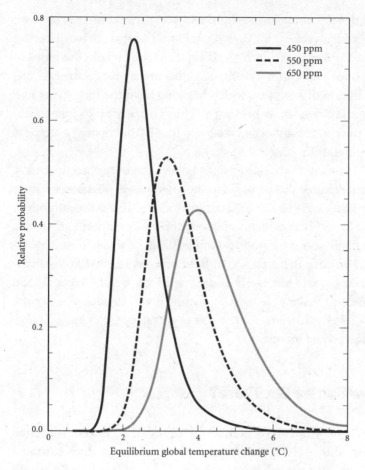

Equilibrium global temperature change (°C)

The likely changes in global temperature (°C) for greenhouse gas stabilization levels of 450, 550, and 650 parts per million. Note that the curves are not symmetric—in each case the tail at the higher end stretches out farther than the one on the lower end. Thus the most probable value for the temperature rise does not coincide with the peak of the curve, but sits slightly to the right. As 1°C is approximately 1.8°F, the likeliest temperature increase for 450 ppm is around 4.5°F (2.5°C), for 550 ppm is 6°F (3.5°C), and for 650 ppm is 7.5°F (4°C). The asymmetry of the curves also means there is a greater risk of temperatures considerably above these "likeliest" values than there is of temperatures that fall below. All in all, these curves show how big a risk we are already taking with our greenhouse future. (Source: The Hadley Centre, based on work published in G. R. Harris et al., "Frequency distributions of transient regional climate change from perturbed physics ensembles of general circulation model simulations," *Climate Dynamics*, vol. 27, no. 4, pp. 357–75, 2006)

But if we're unlucky, even at the lowest greenhouse gas level we can conceivably achieve, we could still find ourselves heading for a highly dangerous 6.5°F. If we allow the greenhouse gases to rise above 450 ppm, the situation becomes even worse. A concentration of 550 ppm, which has long been the target that was supposed to keep us below 3.5°F, has virtually no chance of doing this, and might even take us up to 9°F. That doesn't even bear consideration.

For all of these reasons, we believe the only choice we have is to keep greenhouse gases at the lowest level possible. In other words, we have to aim for 450 ppm CO_2eq. This is the number to look for in every policy statement and every climate agreement. We'll talk about the political implications of this in part 3.

For now, the message to hold on to is this: Although dangerous climate change is already with us, we still have a good chance of being able to cope with it if we stay below 450 ppm CO_2eq. Any higher, and the risk of catastrophic climate change becomes just too great.

How Can We Get There?

Much of what we've said so far sounds like unremitting bad news. But there is still hope. Though it's now almost impossible for us to stay below 3.5°F, this is not a dramatic threshold above which something bad would immediately happen. Rather, it's like a speed limit in that the higher you go above 3.5°F, the greater the risk of a disaster. Everything we've said so far in this chapter should be an extra spur to action rather than a cause for dismay, because we still have a chance of keeping greenhouse gases to that 450 ppm limit.

We will, however, have to act fast. Global greenhouse emissions will need to peak within fifteen years, and by 2050 they will need to have fallen to half their current levels. That sounds

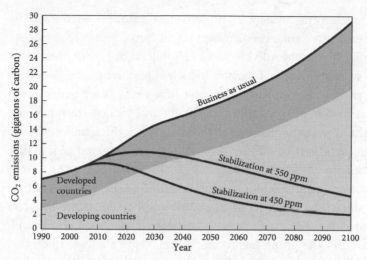

Predicted future greenhouse emissions according to several different scenarios. To convert from gigatons of carbon to gigatons of carbon dioxide (used throughout this book) multiply by 3.67. (Source: UK Department of Environment, Food and Rural Affairs)

like a lot to ask, especially when you consider that much of this change will have to come from developing countries that are currently much more focused on improving the wretched lives of their citizens than worrying about the global climate.

The pace of emissions is also picking up alarmingly. If we don't do anything to change our ways, in 2035 the air will already contain 550 ppm CO_2eq.[7] By the end of the century, emissions could be 250 percent of today's level, with greenhouse gases heading toward the 1,000 ppm mark. The IPCC report describes the challenge of turning this around as "daunting."[8]

However, the good news is that many of the technologies we will need to curb greenhouse gases are already available or in the pipeline. In the next few chapters we will describe these in more detail. But first, just to show you how, together, these technologies can solve the problem, we'd like to mention an ingeniously simple approach put forward in 2004 by two

Princeton researchers, Steve Pacala and Robert Socolow.[9] The researchers broke the problem down into a series of "wedges," each of which would cut global emissions of greenhouse gases by 25 billion tons of carbon (which is just over 90 billion tons of carbon dioxide) over the next fifty years. They called them "wedges" because their effects gradually increase from nothing right now, at the thin end of the wedge, to a thick billion tons of carbon (about 3.7 billion tons of carbon dioxide) per year in fifty years' time.

Pacala and Socolow originally added seven wedges together to show how you could turn a rising graph of emissions into one that stayed level. In fact, as we've shown above, staying level isn't enough—the graph has to fall. By the same analogy, to do this you could add more wedges.

The wedges include the following for fifty years' time (in each case we would need to scale up our efforts in a straight line, starting from zero today):

❑ Double the fuel economy of two billion cars.

❑ Halve the annual average distance traveled by two billion cars.

❑ Cut carbon emissions from buildings and appliances by one-quarter.

❑ Capture and store carbon dioxide from 800 gigawatts of coal power plants.

❑ Capture and store carbon dioxide from 1,600 gigawatts of natural gas power plants.

❑ Build two million 1-megawatt wind turbines (about fifty times what we have today).

❑ Stop all felling of tropical forests and plant 740 million acres of new trees in the tropics.

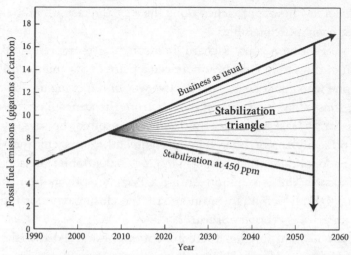

The "wedges" strategy for reducing future greenhouse emissions from the "business as usual" path to stabilization at 450 ppm of carbon dioxide. To convert from gigatons of carbon to gigatons of carbon dioxide (used throughout this book) multiply by 3.67. (Source: adapted from S. Pacala and R. Socolow, "Stabilisation wedges: solving the climate problem for the next 50 years with current technologies," *Science*, vol. 205, pp. 968–72, August 13, 2004)

❑ Double the current amount of nuclear power.

❑ Quadruple the amount of natural gas used to generate electricity by converting coal-fired power stations (since gas produces fewer greenhouse emissions than coal).

❑ Increase the use of biofuels in vehicles to fifty times today's level.

❑ Use low-tillage farming methods on all the world's cropland.

❑ Increase the global area of solar panels by a factor of seven hundred.

You could imagine other wedges put together from the technological solutions that we've described here. Note that none of these requires any major technological breakthrough.[10] If you added a few new approaches along the way, the task would become even more possible.

In February 2007 Richard Branson, the entrepreneurial head of the Virgin Group, announced a prize of $25 million for anyone who could devise an effective way of removing one billion tons or more of carbon dioxide from the atmosphere per year for at least a decade. (Branson was inspired, he says, by the prize of £20,000 offered in the eighteenth century for measuring longitude, which was won by clockmaker John Harrison, and also by the $10 million Ansari X Prize for private human spaceflight, which was won in 2004.) The closing date for the Virgin prize is February 8, 2010.

If you want to enter the competition, be warned that the problem is harder than it sounds, because carbon dioxide is so widely dispersed. There are only four molecules of CO_2 for every ten thousand molecules of air, so any attempt to capture it will also have to find a way to make it much more concentrated.

One promising approach was announced in April 2007 by Klaus Lackner from Columbia University, working with research and development company Global Research Technologies. Lackner's idea is to do the opposite of what chimneys normally do. Instead of pumping CO_2-laden air out into the atmosphere, gigantic chimneys would pump polluted air in from their surroundings, before a mineral soaked up the carbon dioxide and then released it as a pure stream of gas ready for storage underground. An advantage of this approach is that the chimneys wouldn't need to be near the places where the CO_2 was emitted. Instead of being bolted to power stations or trailed behind cars, chimneys such as these could strip the CO_2 out of the air half a world away—for example, right next to a potential site of carbon dioxide storage.

The company estimates that a million such devices, each with an admittedly huge opening of a thousand square feet, would be able to soak up a billion tons of CO_2 a year. If we started very soon with pilot projects, gradually increased our efforts, and ended up in fifty years' time with about four times that capacity, this could make up one of Socolow and Pacala's wedges.[11]

Another way to take carbon dioxide directly back out of the atmosphere would be to let biology do the concentrating for you. Plants build their bodies from carbon dioxide in the air. Normally, this returns to the air when the plant's body is broken down, by being digested in the gut of an animal, say, or burned in a fire. But what if you protected the body so that its carbon stayed safely stored? For instance, planting new trees in the tropics and then protecting them from being logged or burned would be an excellent way to soak up carbon dioxide from the air—which is why it shows up in one of Socolow and Pacala's wedges.

Another intriguing approach that recently emerged is the technology of "biochar." The idea is that you take some kind of agricultural leftover—perhaps the stalks and cobs from corn—and burn it at a very high temperature in the absence of oxygen. Some of the carbon becomes an oil that you can then burn to produce energy. The rest forms a hard, black form of charcoal in which, its proponents believe, it will stay locked up for hundreds of years. The economic benefits of the oil can then help pay for the costs of sequestering the rest of the carbon. One practical side benefit of this approach is that, in principle, the charcoal could then be spread on fields. It wouldn't break down, but it could help the soils to hold on to nutrients and water, which would reduce the need for both artificial fertilizers and irrigation. The idea is still in its infancy, but it could well be one to watch.[12]

One approach we don't recommend is to try to engineer our way directly out of trouble by somehow balancing the warming we have caused with an equal amount of cooling. This idea has a troubled history, partly because many of the schemes proposed over the years have been wildly impractical. Suggestions have ranged from the wacky (paint as much of the world as possible white so that it will reflect more sunlight) to the ultraexpensive (put gigantic mirrors in space that would block sunlight—and would also cost trillions of dollars to make and deploy).

Moreover, this approach, which involves letting the carbon dioxide do its stuff and then wiping out the warming effect, does nothing for the other consequence of high carbon dioxide. An upside could be that the extra carbon dioxide might help to fertilize plant growth without the usual associated droughts that send crop yields tumbling in a warmer world. But a much more serious downside is that, as we explained in chapter 3, putting more carbon dioxide in the air creates a more acidic ocean—and putting a mirror in the sky would do nothing to stop that from happening.

But the main reason that most climate scientists have given the idea of geo-engineering short shrift is that they are suspicious of any quick fix that leaves us with our old bad habits intact. As Meinrat Andreae from the Max Planck Institute for Chemistry in Mainz, Germany, put it: "It's like a junkie figuring out new ways of stealing from his children."[13]

In 2006 atmospheric scientist and Nobel laureate Paul Crutzen made the bold suggestion that we should inject clouds of sulfur dioxide into the air to make an artificial haze—a cloudy mirror in the sky—that would reflect sunlight back out to space and help to balance the greenhouse warming.[14] However, deliberately interfering with something as complex as the atmosphere in this way could be even more dangerous than the problem that action is trying to solve.

The history of our dealings with the atmosphere is littered with unintended (and often unfortunate) consequences. For instance, the chemicals that nearly destroyed the ozone layer—which protects us from cancer-causing ultraviolet rays—were invented by a well-meaning chemist who was trying to improve the environment. Trying to fix the problem by putting yet more pollution in the sky seems perilously like the children's song "I Know an Old Lady Who Swallowed a Fly."

7

MORE FROM LESS

The first and perhaps most important thing we must do to tackle climate change is to get smarter about how we use energy. Fossil fuels have provided us with such a cheap source of power that we've become foolishly profligate with our energy usage. It's amazing how much we waste, without even noticing how much it costs us. In fact, far from being expensive, many of the strategies we could use to reduce carbon dioxide emissions by improving efficiency will actually save money. This is the low-hanging fruit of the emissions game, the gaps that would already have been plugged had the energy market truly been doing its job.

Buildings

Globally, buildings are responsible for greenhouse emissions of about nine billion tons of CO_2eq per year,[1] which is about 18 percent of the total. Most of this comes from energy use, or abuse. Although insulation has improved and appliances have become more efficient over the past few decades, the way we use energy in our buildings is still incredibly wasteful, a combination of pouring greenhouse gases into the atmosphere and money down the drain.

The technologies already exist to put this right. For example, simply using the most efficient new forms of lighting

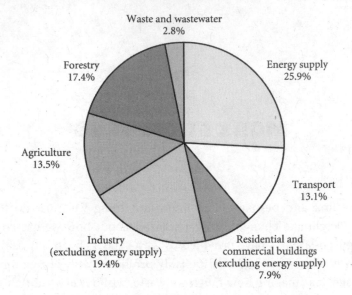

Total greenhouse gas emissions from different sectors. (Source: IPCC)

could reduce the energy used to light a typical American home by up to 75 percent.[2]

Though some of this would cost money to implement, much would come absolutely free, or even save money, thanks to the savings in energy costs. Even in the short term, over the next fifteen years we could cut at least 30 percent from the projected rise in emissions from buildings without paying a penny. That's not counting changing our behavior in buildings, which would enable us to cut even more.

This is one of the many areas of the climate change game where individuals can make a dramatic difference. Many of the following strategies can apply to your own home, place of work, school, or place of worship. If we all start to put some or all of them into place, that will have a huge effect on global emissions.

Insulate and Isolate

It makes clear climate sense to ensure that your home is as well protected from the elements as possible. Any heat that escapes in winter, or cool air in summer, means that you'll have to waste more emissions. If you live in a house, make sure the insulation in the loft is as good as it can be. Seal off drafts around windows and consider lining curtains. And, yes, consider changing the level at which you set your thermostat. Heating and cooling take up more than half the energy use in a typical U.S. home,[3] and most of our homes are overheated in the winter and overcooled in summer. The Department of Energy's Energy Efficiency and Renewable Energy Program has some good ideas about how to make your home more energy efficient.[4] Also, if you're moving, ask the real estate agent or landlord how efficient the new house is, and don't be afraid to demand changes—such as improved insulation—that will make it better.

In warmer regions, many of us have learned to rely on energy-hungry air-conditioning to keep the interiors of our buildings cool. However, another approach is simply to let the building itself act as a filter, selectively letting in daylight, warmth, and cool air, depending on the needs of the moment. If this sounds innovative, it's probably worth mentioning that early architects were doing this before anyone had even heard of oil or electricity. The designers of Moorish palaces in Andalucía, for example, knew all about how to set the slope of a roof so that it blocked the searing overhead summer sun, but let in the gentler, more sloping sunbeams of winter. Modern buildings can also be designed to do this expertly, but in the meantime any of us can work with what we already have. When the sun is shining directly on the windows, close the shutters or draw the blinds.

New Lamps for Old

Lighting is one of the easiest ways to cut the emissions from buildings, which could drop by up to 90 percent using technologies that are already available. For instance, swapping old incandescent lights for the lowest-energy ones available can cut the energy used in half. Designing buildings to make proper use of incoming daylight around the edges can cut the remainder in half again, while a very simple strategy called task/ambient lighting can reduce it still further. This involves having a relatively low background lighting level, with more light at individual workstations. Not only does this take less energy, it's also often more restful on the eyes.

Finally, about one-third of the world's population has no access to any form of modern energy and depends on lighting that burns fossil fuels directly—such as kerosene or paraffin lamps. These provide only 1 percent of the world's lighting but 20 percent of its lighting-related emissions. A compact fluorescent lightbulb is one thousand times more efficient and also creates fewer hazards to health from fumes. This is one example of how, if developing countries can get the funds and knowledge to leapfrog to modern technologies, the whole world will be better off.[5]

Appliance of Science

Most home appliances, such as refrigerators and washing machines, have been getting more efficient with each new design. Manufacturers have done this expressly because governments have required that machines tell you how much energy they use, and customers have voted with their wallets. It makes sense to get the very latest and most efficient machines each time you need new ones. It also makes sense to wash your clothes at the coolest temperature possible.

The Energy Saving Trust points out that electronic gadgets, meanwhile, have been going the opposite direction, getting hungrier for energy with each new design. Digital radios, for instance, draw much more energy than their analog counterparts, and plasma TVs are monster power users. If you want to get a flat TV, go instead for an LCD screen, which is much less energy intensive.

It's good general practice to minimize your gadgets as much as possible and always to request the most energy efficient. This isn't about pinching pennies; it's about treating energy as the precious resource that it is. Also, as with home appliances, the more customers seek energy-efficient gadgets, the more pressure there will be on manufacturers to meet the demand. In 1978 the U.S. government introduced simple federal standards for appliance efficiency. By 2020 this will have cut residential greenhouse emissions by about 10 percent.

Switch to a "Green" Supplier

See whether you can switch your energy supplier to a company that provides only renewable or low-carbon energy. There won't be enough of this on the grid to go around for everybody, but the more people push for it, the greater the pressure will be on electricity companies to find alternatives.

Pull the Plug on Standby

Amazing though it sounds, those little lights on TVs and videos and all the other accoutrements of appliance standby modes really do make a big contribution to greenhouse emissions. It's a testament to our profligacy with energy that even mobile phone chargers that are not in use continue to draw electricity as long as they're plugged in. The International Energy Agency (IEA) estimates that standby mode could be causing a full 1 percent of the world's greenhouse emissions, which is nearly as much as the entire aviation industry.[6]

All that unnecessary electricity use wastes money as well as emissions. According to researchers at the Lawrence Berkeley National Laboratory in California, residential consumers in the United States spend more than $4 *billion* on standby power every year.[7] In 1999 the IEA launched the "one watt initiative" to encourage manufacturers to limit standby mode to one watt or less.[8] In the meantime, though it might sound absurd, it really does make a big difference if you switch every appliance off at the wall as soon as you're no longer using it.

Change Behavior

Even when houses are otherwise identical, one classic study showed that a single family can easily use twice as much electricity as its next-door neighbor.[9] More recently, a survey of fifty research projects worldwide looked at whether or not a household would change its consumption if its members knew how much electricity they were using compared to the same month the previous year, or compared to a neighbor's use. The upshot was an energy saving of 5 percent.[10] This shows that if we're properly informed about how much electricity we're using and pay attention, we're much more likely to use less. It's amazing how a "smart meter" ticking away next to the fridge can encourage you to switch off the lights when you leave a room.

Do It Yourself

Otherwise known as micro-generation, this approach involves using the resources of your building to make your own energy. We'll describe low-carbon methods for generating energy in more detail in chapter 9, but it's important to note that many of these can apply quite well on the small scale. For instance, you can use the sun directly for warming water or generating electricity through solar panels; heat pumps can carry warm or cool air in or out of the building; and individual wind turbines can also play their part. Small-scale "combined heat and power" (CHP) units will also be available soon to make use of both the heat and electricity generated by burning natural gas.

Many of these technologies are currently expensive compared to their fossil-fuel equivalents, but they will become less so as economies of scale begin to kick in. More flexible national grid networks could even enable individual households to export electricity that they don't use back to the grid itself. If it's satisfying to watch a smart meter tell you that you're saving electricity, how much more so would it be if the meter told you that you were selling electricity back to the power companies?

Industry

Overall, industry is responsible for some twelve billion tons of greenhouse gases, almost one-quarter of global emissions. Most of these come from the use of fossil fuels to make energy; direct use of fossil fuels for chemical processing and smelting; or direct production of carbon dioxide as part of the processing (for instance, in making cement from limestone).

Almost all the energy use is concentrated in a handful of energy-intensive industries: iron and steel, chemicals and fertilizers, cement, glass and ceramics, and pulp and paper. There are already plenty of ways to cut down on the energy needed. For example, in heat cascading the waste heat from the highest-energy processes is used to feed a series of successively cooler steps until the final output is used for heating the factory itself.[11] This method would need no new technological advances or new legislation, and many factories could employ it tomorrow.

Another technology that is already available to improve efficiency is variable speed motors.[12] More than half of all electricity generated is used to drive motors, and increasing their efficiency could make a huge difference. A motor that operates continuously and flexibly at relatively low speeds uses much less power than one that turns on and off at a fixed high speed. Similarly flexible motors on fans could save up to one-third of their energy, and elevators could regenerate energy when going down.

Industries could also use the heat generated by electric power plants, which is usually just wasted. If a factory were sited close enough to a power plant, the heat that normally disappears into the air via the cooling towers could instead be piped in to feed the heat-hungry processes. (This approach could also deliver heat to individual towns through district pipe networks.)

For the industries that generate carbon dioxide directly—such as the cement industry, which is currently responsible for some 3 percent of global greenhouse emissions—the most promising approach would be to capture the carbon and store it before it can reach the atmosphere.

Agriculture

Agriculture accounts for about 13 percent of global greenhouse gas emissions, approximately the same amount as transport. Almost none of this comes from carbon dioxide. Though huge amounts of CO_2 do pass between the atmosphere and agricultural crops every year, the balance is more or less zero. It's the sister greenhouse gases—methane ("natural gas") and nitrous oxide ("laughing gas")—that matter most here.

Agricultural methane comes from a variety of places, and almost always involves microbes feasting on organic matter in places where there is little or no oxygen. Thus the biggest sources are the guts of cows, sheep, and water buffalo. A smaller amount of methane comes from rice paddy fields and from the burning of biomass. (Biomass burning, especially of forests, also contributes significant amounts of carbon dioxide, but we'll deal with this problem separately.)

Nitrous oxide comes directly from soils, especially when they have a surfeit of nitrogen-containing compounds because they have been doused with nitrate fertilizers or with animal manure.

Nobody has yet come up with a detailed study of exactly how these emissions will increase in the future if nothing is done to stop them. However, the IPCC report estimates[13] that by 2030 the agricultural contribution to these two greenhouse gases will have risen by 31 to 37 percent.

There are various ways to control agricultural emissions, almost all of which involve increasing the efficiency of how we use our land, and the technologies by and large already exist. For instance, better diets for livestock makes them—to put it frankly—belch less. In New Zealand, scientists are studying how to change the microbes that cows use to digest their food to encourage them to make sugars instead of methane. Even independently of climate change this makes good agricultural sense, since any loss of carbon by methane means less carbon to add to the bulk of the animal's body. It's also possible to control the nutrients in the soils by more natural means than adding nitrate fertilizers. For instance, the practice of agroforestry involves planting crops that contribute to each other's nutritional needs side by side. Agricultural lands could even be encouraged to become net sinks of carbon by reducing the amount of plowing, which disturbs the soil and encourages microbes to mobilize the carbon that it contains.

None of these practices would necessarily come cheap, or be immediately practical for the subsistence farmers in developing countries who will contribute the bulk of the growth in "business as usual" greenhouse emissions. But if we come up with a way to make emissions cost money, by putting a decent price on the heads of all the greenhouse gases, this could help to pay for the agricultural changes we need. The IPCC report says that with a price up to $50 per ton of CO_2eq, total agricultural emissions in 2030 could be reduced by about one-third, and with a price up to $100 per ton they could be cut in half—which would more than wipe out all the projected increases.[14]

Forests

Chopping down and burning trees contributes some eight billion tons of carbon dioxide to the atmosphere each year.[15] That's a huge amount, more than 16 percent of the total human greenhouse gas emissions—more than comes from either agriculture or transportation. And yet it's largely unnecessary. Destroying forests is one of the maddest forms of interfering with the climate that we humans have yet devised.

Most of the carbon dioxide from deforestation comes from the burning of tropical forests, especially the Amazon rain forest. Reasons for cutting it down range from logging of individual hardwood trees to slash-and-burn for subsistence agriculture, to clearing land for large-scale agricultural plantations of palm oil. Even logging of individual trees often creates mass destruction. Bulldozers blast through the remaining trees to get to their prize; felling of individual trees often brings down their neighbors, as they are bound together with woody lianas; the holes thus created in the leaf canopy let in the hot tropical sunlight, which dries out the forest and allows for the outbreak of accidental fire; and loggers usually build roads, which encourage the influx of subsistence slash-and-burn farmers.[16]

Models suggest that by 2030 a combination of preventing current deforestation, better management of forests, and reforesting appropriate areas[17] could easily and cheaply reduce annual emissions by up to fourteen billion tons per year.[18] Few studies have made predictions as to what the baseline "business as usual" scenario would be for this time period, though there is some suggestion that deforestation emissions would be more or less the same per year in 2030 as they are today—in other words, eight billion tons per year. In that case, applying all these tactics would mean that not only would we reduce forest emissions to zero, but forests would actually be soaking up considerable amounts of the emissions that came from other sources, helping us to meet the overall global reductions that we need.[19]

There would also be very little loss of food production. The soil that supports tropical forests is relatively poor in nutrients, most of which are bound up in trees, leaf litter, and endless biological recycling. Thus, it is inadequate for growing crops for more than a few years, which is one reason why slash-and-burn farmers need to keep moving. In central Brazil advances in soil science have allowed farmers to grow crops such as soybeans in the previously infertile Cerrado region, which has taken pressure off the Amazon while the overall agricultural production has actually gone up.[20]

What's more, any reduction in emissions from forests would be relatively cheap, especially when compared to other ways of cutting carbon. Economic research suggests that the direct yield from forest land that's been converted to farming, including the proceeds from the sale of timber, is the equivalent of less than $1 per ton of carbon dioxide in many areas, and usually well below $5 per ton.[21] Any economic system that makes it expensive to emit carbon dioxide by putting a price on the head of every ton of CO_2 would wipe out these profits many times over. Analyses suggest that almost half of the reduction would cost less than

$20 per ton of CO_2, and more than two-thirds would cost less than $50 per ton.[22] As we'll show in chapter 10, these numbers look eminently sensible when compared to the likely prices that will be set in future carbon markets.

How to ensure that the countries where deforestation rates are highest have both the funds and the incentive to put a stop to it is a political rather than a technological question; chapters 10 and 11 discuss some of the strategies that are already in place, as well as ideas for how to strengthen them. What's clear is that we need to make it happen. It's true that many developed nations, especially in Europe and North America, converted their forests to agricultural land centuries ago, and that this laid the foundations for their further development. That's one reason why the developed world should help pay to reduce tropical deforestation. To those who exploit them, tropical forests might seem like a cheap and almost infinite resource. But the emissions they are putting out as they burn are costing all of us the earth.

Waste

Waste is a relatively small contributor to emissions at around 1.2 billion tons CO_2eq per year, which is just a few percent of the global total. Still, it's also easily reduced. Most of the emissions come in the form of methane from landfill sites, which is relatively straightforward to capture and can often be used to generate electricity. Even more important, though, are the indirect emissions from waste, which are much harder to quantify but likely to be very substantial. From a climate point of view it's foolish to use things once and then throw them away. The archaeologist E. W. Haury wrote: "Whichever way one views the mounds [of waste], as garbage piles to avoid, or as symbols of a way of life, they . . . are the features more productive of information than any others."[23] If alien archaeologists began to uncover our own particular piles of plastic bags that have been used only once and then thrown away, along with the wrappings, cartons, tins, and papers that have served us very little but cost us dear in carbon emissions to make, it's not hard to imagine what conclusions they might draw.

8

PLANES, TRAINS, AND AUTOMOBILES

Transportation has received so much attention in the climate change story that we have given it a chapter of its own. This sector currently makes up 13 percent of global greenhouse emissions, and after power generation it is by far the fastest growing.[1] It's also by far the most talked about, probably because it's the one area in which individuals can make the most dramatic difference. Almost all the energy for transportation comes from petroleum, and unless there is a dramatic change from our current path, greenhouse emissions from transportation will be 80 percent higher in 2030 than they are today.[2]

Part of the problem is that wealth and carbon-hungry transportation have long tended to go hand in hand. The richer the world gets, the farther it wants to travel and the less carbon efficient the means it chooses. As the wealth of a nation rises, its citizens change from walking and cycling, through buses and local trains, to cars, high-speed trains, and air travel.[3]

Most of the projected rise over the next few decades will come from developing countries, but it's hard to argue that the citizens of those countries should not enjoy the advantages of travel already afforded to those in the industrialized world. Curbing these emissions will take a combination of new low-carbon technologies, increased efficiencies, and the provision of attractive enough alternatives to encourage us all to turn readily away from the most polluting methods of moving around.

Biofuels

First, a few words of warning about one of the highest-profile new technologies for transportation—low-carbon fuels, known collectively as biofuels. Though most of the strategies for low-carbon energy produce electricity, transportation almost always needs a liquid fuel that can easily be delivered to individual cars, buses, trains, and planes. (Electric cars are one promising exception to this, as we discuss below.) One way to provide a liquid fuel without damaging the atmosphere would be to make new green fuels that use carbon-neutral biomass as a feedstuff. That way, we could keep doing more or less what we do now, but with fewer carbon consequences.

This is such an attractive notion that two of these biofuels are already in use: bioethanol, which is a direct substitute for gasoline, and biodiesel, which is just what it sounds like. Bioethanol can be made from more or less any crop containing sugar or starch. The starchy crops, such as wheat or potatoes, would then need an additional chemical process to convert their starch into sugar. Yeast is then added to the sugar and the mixture fermented, as for making alcohol to drink. Similarly, biodiesel comes from crops that make oil, such as rapeseed, sunflowers, or oil palms.

In principle, biofuels like these should be carbon neutral because the emissions produced by burning the fuel balance the carbon dioxide taken up by the crops from the air in the first place. But as with so many other attempts to replace fossil fuels, the environmental devil is in the details.

First, there is the danger that agricultural land can be diverted from food production, or that grains that would have been used for food are instead used to make fuel. That in turn would drive prices for existing foodstuffs higher, which hurts those least able to afford to feed themselves.

This is already happening. In June 2007 the United Nations Food and Agriculture Organization published a report warning that the demand for grains and vegetable oils to convert into biofuels was already driving prices very high. The report estimated that food import prices for developing countries have risen by a shocking 90 percent since 2000.[4] Considering that many of those same developing countries are already suffering the effects of climate change brought on them by the West's use of fossil fuels, this attempt to right the wrong looks more like heaping injury on injury.

Growing the necessary crops also means using nitrogen fertilizers, which in turn cause the soil to put out nitrous oxide, a potent greenhouse gas in its own right. Then plowing, harvesting, and processing the crops require energy, which is usually supplied by burning fossil fuels. Each extra step—converting starch to sugar, converting sugar to alcohol—takes more energy still. There's even the possibility that certain biofuels will produce more greenhouse gas emissions than the old fossil fuels they seek to replace.

President Bush has sought to boost his climate change credentials, and the prospect of improved fuel security, by spearheading a rush to make bioethanol from corn. However, the evidence suggests that this could hurt as much as it helps. Many researchers have tried to do a life-cycle analysis of bioethanol from different crops, but it's still hard to know exactly what numbers to use in the calculations. For instance, a group of scientists from Berkeley, California, reviewed a whole set of studies and concluded that bioethanol made from corn produced some 18 percent fewer greenhouse emissions than normal gasoline. Though it's nowhere near 100 percent, that might sound promising—until, that is, you notice the uncertainty. Though 18 percent is the researchers' best guess, they still acknowledge that the real number could be anywhere from 36 percent less than gasoline (which would be better still) to

29 percent *more* than gasoline (which would obviously be much worse).[5]

A report from the International Energy Agency published in 2007 concluded that the typical carbon dioxide reduction for biofuels based on corn was 15 to 25 percent compared with gasoline. Using sugarcane as a feedstock, on the other hand, produced up to 90 percent CO_2 reduction and should be considered a much more viable alternative.[6]

It's also important to calculate how much carbon dioxide biodiesel costs to produce. In addition to the processing costs, there's the question of where in the world the vegetable oil to make it comes from. The European Union has committed all its member states to blending fuels with 5.75 percent biofuels by 2010, which has sparked a new oil rush. Since Europe doesn't have enough land to supply the necessary vegetable oils, it will have to import on a huge scale. This is good news for countries like Malaysia and Indonesia, which produce most of the world's palm oil. But environmental groups warn that these countries are planning to cut down large tracts of virgin rain forest to convert to palm oil plantations.

Replacing a vivid, varied ecosystem with regimented rows of a single species of palm isn't just an aesthetic loss (though it could be the final blow for that charismatic primate the orangutan); it's also dangerous. The sheer variety of species in a rain forest protects it from incoming diseases, whereas a disease that specifically targets oil palms could wipe out an entire plantation almost overnight.

Then, of course, there's the issue of carbon dioxide. Burning vast amounts of virgin rain forest, together with the potential of such fires to spread to the region's carbon-rich peat lands, could put enough emissions into the atmosphere to outweigh the carbon benefits of biodiesel for decades. Such so-called green fuel would therefore be anything but.[7] A United Nations report published in 2007 was careful to state that, though most biofuels provide fewer carbon emissions than their fossil-fuel rivals, this is only "provided that there is no clearing of forestland or . . . draining of peat lands that store centuries of carbon."[8]

All this is not to say that biofuels are bad. In fact, they can work very well. For the past thirty years they have been a superb resource for Brazil, which responded to the 1970s oil crisis by converting sugarcane to bioethanol and now produces over half the world's supply. The scheme has worked well there at least in part because the country has large tracts of well-watered agricultural land, which is perfect for sugarcane production. Brazil is now sharing its expertise with countries in southern Africa, to enable them to resuscitate their moribund sugarcane industry and put it to biofuels use.

The important point to make is that not all biofuels are equal in the fight against climate change. The UN report suggested a way around the problem of rogue emissions during the planting and processing, in the form of an international certification scheme to state exactly how much greenhouse gas the bioenergy has saved, or indeed cost, throughout its lifetime.[9] "While developing and implementing a widely accepted certification scheme will be a challenge," it adds sternly, "this should not deter governments, industry and other actors from making the effort."

The best way around the problem of competition with food would be to find a way to use cellulose, the material that lines the cell walls of plants and makes them stand tall. Unlike horses and cows, humans can't digest cellulose, but it is packed with sugars. If these sugars could be converted into ethanol, this would open up the possibility of using stalks, wood chips, and tough wild grasses that would never compete with food crops. Termites have the technology to do this. They secrete an enzyme that helps them digest cellulose for food and we could do well to learn from their example. Eight small-scale factories are currently under construction in the United States to convert crop leftovers into ethanol, in Iowa, New York, Tennessee, Michigan, Georgia, Kansas, Colorado, and Florida. However, cellulose conversion will require much more research and development before it can truly compete.

Planes

Airplanes have become the new villains of climate. In terms of overall numbers, that's not strictly fair. Aviation is directly responsible for about seven hundred million tons of carbon dioxide each year, which is only 1.6 percent of global greenhouse emissions.[10] However, molecule for molecule the emissions count for much more than they would on the ground, because planes are very efficient at causing greenhouse warming. High-altitude deliveries of nitrogen oxides (which form ozone, another greenhouse gas), as well as the water in contrails that can go on to form cirrus clouds, together enhance the direct effect of carbon dioxide by up to a factor of three.[11]

There's also the inequality argument. One reason why planes contribute so little to the overall greenhouse gas burden is that relatively few among the human population can afford to use them. A round-trip flight from Los Angeles to New York, for example, generates a hefty 1.5 tons of carbon dioxide per passenger. Without even counting the extra greenhouse effects involved in flying, that's as much as an average citizen in India emits in an entire year. (To be fair to aviation, this argument also applies to power generation. As we've already mentioned, fully one-third of the world's population has no access to modern energy sources.)

The third reason why aircraft receive so much attention is that, unlike the case with other carbon-hungry forms of transportation, there are no likely alternatives to fossil fuels on the horizon. Hydrogen could work, and would be especially low-carbon if it was made from renewable energy sources rather than fossil fuels.

However, it would probably have to be carried on the plane in liquid form, which would be much heavier than the kerosene that airplanes now use and would also need to be refrigerated. Worse still, the output from burning hydrogen is water. That's fine on the ground, but up in the air it would contribute to making further cirrus clouds, which would in turn help to warm the planet.

Another possibility would be to make kerosene artificially. This process is already well established starting from coal, or in some cases natural gas. In principle it could also be done using biomass, such as the waste left over from food crops, for a low-carbon alternative. The technology merely needs to be adapted, but it's not clear whether it would become competitive given the relative cheapness of pulling the feedstuff for kerosene out of holes in the ground.

Biofuels are perhaps more promising. However, ethanol yields less energy, pound for pound, so tanks would have to be much larger to accommodate fuel for the same distance traveled. And biodiesel freezes at the low temperatures encountered aloft, so the fuel tanks and engines would have to be heated.

One alternative might be butanol, which is like ethanol but with twice as many carbon atoms in its chemical backbone. Because of this, butanol packs more energy punch for its weight and also freezes at much lower temperatures. (Furthermore, it's less corrosive than ethanol, so would be much easier to transport in pipelines.) In 2007 Virgin Atlantic and Boeing announced a joint initiative to work with engine manufacturer General Electric Aviation on developing a plane that could fly on biofuels. The plan is to test-fly one of Virgin's Boeing 747s sometime in 2008, with a mix of biofuel in one of the four engines.

The head of Virgin, Richard Branson, suspects that butanol might be the answer, though he's not ruling out finding something else that will work even better. In 2006 Virgin announced that profits from its flights and trains over the next ten years (more than $1 billion) will be plowed into a new venture, Virgin

Fuels, which is dedicated to finding new low-carbon fuels for transport. This seems to be sound business sense rather than well-meaning philanthropy. Whoever invents an effective low-carbon fuel system for airplanes will be able to look Bill Gates in the bank balance.

Oil giant BP is also betting on butanol, though not necessarily for planes. In 2006 it announced that it was investing $1 billion to develop new biofuels, and other large energy companies—as well as the U.S. Department of Energy—are also starting to commit funds.

Even if the biofuels experiments pay off, they're unlikely to replace standard jet fuel for several decades.[12] So for now, most of the efforts to reduce the environmental effects of aircraft depend on improving their fuel efficiency. Partly because jet fuel already makes up a high proportion of the cost of flying, aircraft manufacturers have a good record for improving efficiency. Modern planes are some 70 percent more fuel efficient than those of forty years ago.[13] The new Boeing 787 "Dreamliner" uses lightweight composites for its body and is 20 percent more efficient again.

Research into other fuel-saving technologies, which was abandoned in the 1990s when oil prices plummeted, are re-entering the arena. For instance, suction systems to reduce turbulence, new body and wing designs, and rear-mounted, open turboprop engines could all improve fuel efficiency by up to a further 60 percent.[14]

Changing the way planes fly would also help. Plenty of fuel is wasted by taxiing and by endless waiting in holding patterns. Several airlines are looking into the possibility of towing planes to the runway and not switching on the engines until just before takeoff. Better air traffic control could also ensure that planes fly the minimum possible distance and land immediately on arrival.

However, even if all these strategies for increased efficiency come into play, they are still likely to be swamped by yet bigger increases in passenger numbers. According to the IPCC report, a "business as usual" scenario, which incorporates projected improvements in efficiency, would produce 2.5 times the current aviation emissions by 2030. Even with more drastic improved efficiency, the emissions would still almost double.[15]

Thus, until a new fully green jet fuel can come into play, the most immediate strategy will have to be to coax people out of planes, especially in cases where other, less carbon-hungry forms of travel are readily available. In particular, short-haul flights produce disproportionate emissions, because so much of the flight is taken up by taxiing, waiting, climbing, and approaching. The message here is that if you can take a train, do so.[16]

If you do fly, try to choose an airline with a relatively young, and hence energy-efficient, fleet. Consider offsetting the emissions from your flight, and, if you do this, make sure you choose a reliable company with projects that adhere to the Gold Standard. (For more on this, see the discussion in chapter 14.)

Meanwhile, it's worth making a point here about international shipping, which has so far escaped the accusing fingers that are being pointed at aviation. Most international freight travels by sea, and ships are responsible for about eight hundred million tons of carbon dioxide per year,[17] which is slightly more than from aviation. Like air travel, shipping is growing at an extraordinary rate. Also like air travel, questions about which country is ultimately responsible for the emissions are so contentious that it is not even included in national emissions tallies. There's more in chapter 11 about how to resolve this issue.

Trains

Trains are already a relatively carbon-friendly way of getting around even using standard fossil fuels. Emissions can be as little as 10 percent of that of the equivalent journey by car, depending on how many passengers are in the car and what fuel the train is using.[18] Train travel can also be much more pleasant than the combination of endless security lines and limited room that constitutes modern air travel, so there's great potential for convincing us to switch. Even for shorter journeys, trains could be made attractive given the right pricing incentives and timetables. This will require considerable investment outside certain areas; while the United States' northeast corridor is well served by trains, much of the rest of the country has as yet little to choose from.

In places where train services are already plentiful, there are also ways to improve them further. Using a low-carbon fuel such as biodiesel would cut carbon emissions to almost nothing, as would electrifying trains and taking advantage of new low-carbon technologies as they come on to the grid. Fueled by France's electricity, which is 80 percent nuclear, the famous TGV train is one of the lowest-carbon ways to travel in the world, short of cycling or walking. If you're visiting Europe and want to travel to almost anywhere in France, this is your best carbon bet.

Coordinating the ticketing and use of trains over long distances and between countries and extending the coverage of high-speed networks will all be important if trains are to compete effectively with short-haul planes. In July 2007 a new Europe-wide network called Railteam was launched, which will soon allow people to book a single ticket for travel across several different countries, and coordinate connections.[19] Though elsewhere in the world transnational train networks are not as advanced, the drive to reduce emissions, especially from freight, might even breathe new life into the long-awaited Trans-Asia Railway, with the aim of coordinating gauges and prices and connecting European countries directly with a broad swathe of Asia.

Automobiles

Three-quarters of all transportation emissions come from road vehicles, and if nothing acts to intervene there will be more than a billion on the roads by 2030, and a billion more again by 2050.

To solve this impending transportation emissions crisis we will need to find some alternative way to drive that does not also produce carbon dioxide. An approach already on the market is hybrid vehicles. Present hybrids, of which the Toyota Prius is the market leader, combine a standard gasoline-fueled internal combustion engine with an electric motor and battery. The idea is that the battery is recharged when the engine is running, and especially during braking. Such hybrids use the electric motor/ battery to boost power during acceleration, to run power steering and other accessories, and to drive the engine at low loads when normal internal combustion engines are at their least efficient.

Under standard U.S. driving conditions, this improves fuel efficiency by up to 50 percent, with an equivalent drop in greenhouse gas emissions. Hybrids might do even better than this in heavily congested urban settings, when the battery can gain proportionally more from braking and when the electric motor can be used proportionally more for slow, otherwise inefficient driving. They also have the general advantage that they can be used with the current gasoline distribution systems, allowing manufacturers to produce cars that are barely distinguishable in range, power, and size from their fossil-fuel counterparts. In 2005 worldwide sales of hybrid vehicles totaled more than half a million. (While that's impressive, it's still only a tiny fraction of the sixty-three million motor vehicles sold that year.[20])

Better still, from a greenhouse perspective, would be all-electric vehicles that could be powered directly from the electricity grid. That way, as more renewable and low-carbon energy made it to the grid, transportation could gain the low-emission benefits. In fact, electricity powered some of the first motorized vehicles until they were outcompeted by the cheap, easily available fossil fuels that powered the internal combustion engine. A few modern electric cars exist, but they suffer from short driving range and very limited battery life. "Plug-in" hybrids might be a good intermediate step. In 2007 both General Motors and Ford announced that they were starting development of hybrid vehicles that could travel short distances without using gasoline at all and be easily recharged at a regular electricity socket. This would be particularly useful if the electricity came from a low-carbon source.

Another important strategy is to improve the fuel efficiency of the cars themselves. More aerodynamic designs, using lighter materials and increasing the efficiency of engines, will help to reduce emissions—though there is always the danger that such innovations will simply provide an excuse to make bigger, more powerful cars. This is another area in which consumer choice is likely to have a considerable effect on the future of cars. If we as consumers insist on cars that are lightweight and economical rather than pointlessly large and power-hungry, manufacturers will have to respond to our demands. If, on the other hand, we stick with the tired old notion that big is beautiful, manufacturers will inevitably do the same.

Something else to look out for is the amount of emissions your car produced while being created. There are now moves to design cars whose "cradle to grave" emissions are as low as possible. For instance, they could be built from lightweight, strong, biodegradable materials prepared at low temperatures, unlike stainless steel.

We will also need to change the way we use our cars. One promising approach is eco-driving, which encourages drivers to use their cars with the maximum possible efficiency. Techniques include smoother deceleration and acceleration, keeping

engine revolutions low, turning off the engine when idling, reducing maximum speeds, and ensuring that tires are properly
inflated. Studies in the United States and Europe suggest that
such measures could improve fuel economy by an extraordinary
20 percent, and in the Netherlands eco-driving is part of the
standard driving-school curriculum. Some cars are also already
fitted with onboard technology aids to get the most out of every
gallon of fuel.

Still better would be making it attractive to people to be less
reliant on cars and to use less carbon-intense forms of transportation, especially for short journeys. Cars consume more energy
and emit more greenhouse gases per passenger mile than any
other surface form of transportation. Almost anything would
be better. Cutting down on the number of cars on the roads
would also bring us all many side benefits: Imagine living in a
world where the roads were safer and where there were fewer
traffic jams, less air pollution, and less vulnerability to erratic oil
prices.

However, history shows that most of us are unlikely to
change our ways unless the alternatives are both readily available and more appealing. This could be especially important for
short trips. In Bogotá in 1998, for example, 70 percent of private
car trips were for less than two miles. The city now has a project,
funded by the Clean Development Mechanism, using buses that
travel frequently along dedicated bus lanes, which makes them
both quick and easy to use. This is projected to save the city
more than seven million tons of carbon dioxide over the next
thirty years, at a cost of less than $20 per ton.[21]

Another possibility in cities where public transportation is
already advanced is to use Internet and even wireless technology to make it more reliable. Seattle's "BusMonster" Web site
has already begun to offer locals a mix of maps and real-time
information on the whereabouts of their buses. Before you leave
home you can check the Internet to find out exactly when your

bus will arrive at the nearest stop, and how long—given current traffic conditions—it will take to get to your destination. In the future this sort of information could be sent directly to your mobile phone and be readily available at every bus stop.

In rural areas, "on-command" buses could soon rival the attraction of the car. In 1999 in the United Kingdom the "Wigglybus" began operating in Wiltshire. It has a core circular route, but if you phone in a reservation it will also make a detour to pick you up. The bus is satellite-tracked and the call center sends a list of pickups to the driver. There are now three separate routes in operation, where rural passengers can make a call and be picked up shortly afterward, often outside their front doors.

Integrating transit with efficient city design will be especially important in the most rapidly developing areas of the world. The populations will be so great that even a relatively small percentage shift from cars to public transportation can make a big difference. One study suggests that strategies like these could almost halve the projected vehicle emissions in a rapidly developing city like Shanghai by 2020.

In the longer term, low-carbon fuel might be the answer. One approach that has received plenty of attention is using hydrogen. The idea here is to use some low-emission energy source to split water and make hydrogen gas. Hydrogen could also be made from natural gas. Although this generates carbon dioxide emissions, it is still more carbon-efficient than burning regular gasoline. And if the carbon could be captured and stored, the emissions could drop almost to zero.

In most versions of this vision, the hydrogen is stored, transported, and delivered to a device known as a fuel cell, which generates power when called on to do so.[22] Fuel cells are highly efficient and are silent when operating. In fact, cars running on fuel cells would be so quiet that it might even be necessary to manufacture noise artificially to warn pedestrians of their presence. But the beauty of this approach is that all that comes out of the car's exhaust is water.

The technology for this is reasonably advanced, and a handful of test vehicles are now on the road. Iceland opened the world's first hydrogen fueling station in April 2003 and is hoping to become a fully hydrogen economy by 2030, with the hydrogen created using the country's carbon-free geothermal energy. Norway's first hydrogen station opened in August 2006, and the government there has plans for a 360-mile hydrogen corridor between Oslo and Stavanger. The European Union is testing hydrogen buses, and China plans to use some at the Beijing Olympics in 2008. Members of the California Fuel Cell Partnership have put nearly two hundred test vehicles on the road, and the state has plans to introduce hydrogen in twenty-one interstate highway filling stations.[23]

At present, though, the technology is still some way from being fully commercial. The main problems are finding the right way to store and distribute the hydrogen, as well as improving the efficiency of the fuel cells and bringing their costs down. However, many large car manufacturers, including General Motors, Chrysler, Mercedes, BMW, and Toyota, are all now working very hard to resolve these issues, and Toyota has provided prototype fuel-cell cars to all Japanese government ministers. When the technology develops to the point of mass production, which could be before 2020, the necessary changes to the existing gasoline-based infrastructure should become commercially viable.

Another way to reduce emissions from the road might be to use biofuels. Most gasoline-powered cars can run with up to 10 percent ethanol in their fuel mix without any modification, and some new cars can burn the pure stuff. The technology was introduced in Brazil in the 1970s, and the flexi-fuel adaptation allows cars to be run on gasoline, alcohol, or any mixture of the two. Biodiesel can also be used with relatively little modification in both cars and trains. Many countries now have targets for the proportion of biofuels to be used in road transportation, though it's vital to know where the biofuels have come from and how many emissions have been produced during the unseen parts of their life.

9

POWER TO CHANGE

Our enthusiasm for burning fossil fuels has scarcely dimmed since we first came up with the idea during the Industrial Revolution. The last three decades of the twentieth century in particular saw a dramatic growth spurt of 70 percent in human emissions of greenhouse gases, so that by 2004 we were putting out a grand total of forty-nine billion tons of CO_2eq per year.[1]

Those emissions are projected to continue their rise for the foreseeable future. Between them, coal, oil, and gas make up almost all our current energy use. There's enough oil and natural gas in the world to last us for decades, and enough coal for centuries.[2]

By far the largest rise since 1970 came from power generation, which increased by nearly 150 percent and now accounts for about one-quarter of our global greenhouse emissions.[3] One way to reduce these emissions will be to make existing fossil-fuel plants more efficient. For instance, switching from the worst fossil fuel in terms of emissions (coal) to the best (natural gas) typically reduces emissions by 50 percent.[4] However, if we are to have any chance of staving off the most dangerous effects of climate change, this won't be enough. We will have to find new ways of making power that don't also involve making greenhouse gases.

Unfortunately, although the technological cupboard is not exactly bare, it's not replete with alternatives, either. In spite of the increasing interest in climate change over the past few years, very little serious industrial attention has been paid to alternative energy sources. Comparing the amount of research and development spent by an industry to its total turnover tells a woeful story. In the United States, the figure for research into power is only 0.5 percent of turnover, compared to 3.3 percent for the car industry, 8 percent for electronics, and 15 percent for pharmaceuticals.[5] Worldwide, investment in energy research and development has remained roughly constant over the past couple of decades, and in some countries—including the United Kingdom—it has even gone down, as oil prices have fallen and deregulation of energy companies has led them to focus more on short-term gains than on long-term research.

But the world is finally starting to wake up to the technological challenge, and efforts are now beginning within the energy industries and in collaboration among them. In the United Kingdom for instance, one of us (David King) devised and set up the Energy Technologies Institute, which will soon be opening its doors for business. A partnership between the government and a consortium of industries, it will receive £1 billion over ten years to focus on developing new, low-carbon forms of energy.

The International Energy Agency predicts that more than $20 trillion will need to be invested between now and 2030 to meet the world's growing hunger for power.[6] Governments and businesses around the world will soon be making decisions about where their investment should go. Following are the low-carbon technologies that will need to be on the world's shopping list to meet this demand.[7] (It's worth mentioning that many of these technologies come with their own side benefits, especially reducing local and regional air pollution.)

Hydroenergy

Hydroelectric power is currently the biggest of all the renewable energy supplies.[8] In 2004 it provided 5 percent of all global energy and a full 90 percent of electricity from renewable sources. However, flooding land to make new reservoirs for hydropower often comes with its own environmental cost. For instance, there is often a loss of ecosystem services. Many of the nutrients that once fertilized the land following the annual flooding of the Nile River now end up instead on the floor of Egypt's giant Aswan Dam, and for the first time in the country's very long history Egyptian farmers are having to put artificial fertilizer on their fields. Flooding any region to create a dam also usually entails the social cost of forced resettlement.

Moreover, there are still some question marks about exactly how much carbon emissions a hydroelectric plant can save. Though the electricity generation itself doesn't cause any greenhouse emissions, shallow reservoirs can give off methane, which is a much more potent greenhouse gas than carbon dioxide. On balance, however, most hydro plants rate much better for greenhouse emissions than their fossil-fuel equivalents.

Because of this uncertainty over the environmental benefits of large hydroelectric plants, the UN has excluded them from its Clean Development Mechanism, through which industrialized countries can fund low-carbon technologies in the developed world. (Chapter 10 has much more about the Clean Development Mechanism and how it works.) This blanket ban doesn't seem the best position to take. We believe it would be considerably better to judge potential projects individually, by doing a full environmental and greenhouse gas analysis, and only then deciding whether it is worth sponsoring from both an environmental and climate point of view.

Geothermal Energy

Using naturally hot rocks to heat up steam and make electricity is clearly a good idea. The only problem with geothermal energy is that you need to be lucky enough to have the hot rocks in the first place. Iceland, which stands over a series of volcanoes, needs merely to drill into the ground to find the heat it needs. But for most other places, there isn't enough heat belowground to generate a significant amount of electricity.

Geothermal heat can still be used to warm buildings directly, using heat pumps. The electricity it takes to pump the heat up from belowground is more than balanced by the savings over more conventional, fossil-fuel-powered heating sources. Many areas are suitable for this, and such heat pumps should become a standard feature of homes throughout the world. New urban developments are best, since it's easier to drill the required bore holes before you build the houses.

Still, geothermal energy presently makes up only 0.4 percent of the overall power generation mix, and it's unlikely to grow significantly in the future.[9]

Bioenergy

This involves generating heat and/or electricity by burning more or less any biological material (also called biomass). That could include logs of wood, wood chippings, straw, agricultural leavings (such as stalks and leaves), fast-growing trees, dedicated crops grown especially for their energy properties, and even the solid waste from sewage or the methane from landfill sites. Globally, bioenergy makes up an impressive 10 percent of the energy supply, though most of this is in simple (and inefficient) wood fires in the developing world, the smoke from which leads to significant health problems. The IPCC report estimates that every year in India alone, indoor air pollution kills half a million women and children.[10]

To make a big dent in the greenhouse problem, biomass would need to be used to replace fossil fuels for large-scale electricity and heating. One problem with this is that biomass is not a particularly concentrated form of energy, especially compared to fossil fuels, whose energy was compressed over hundreds of millions of years. Still, it's already being used as an additive to coal in more than 150 coal-fired power plants around the world.

In the longer term, biotechnological research might produce crops that can be grown specifically for their high-energy characteristics, though as yet there are no such varieties on the horizon.

As discussed in chapter 8, crops could also be a useful source of low-carbon transport fuel.

Nuclear Power (Fission)

Nuclear fission power comes from the energy that spills out when you split heavy atoms. Long the bête noire of environmental groups, it has begun to enjoy a green renaissance thanks to its relatively low greenhouse emissions. In principle, nuclear power should generate no carbon emissions, because it doesn't involve burning any fossil fuels. However, the processing of uranium ore to make the feedstock takes energy, as does the building and eventual decommissioning of the plant itself, and at least for now this is usually supplied by fossil sources.

All detailed studies agree, however, that nuclear power is a very low-carbon energy source and hence a good candidate for the new energy mix. In particular it is one of the very few technologies that are currently ready to go. It's not renewable, in that the uranium ore has to be mined and will eventually run out. But the current stocks should be enough to last for several centuries even if the industry expands significantly, and technological developments in fuel recycling could extend this to several thousand years.[11]

More problematic are the issues of waste, safety, and nuclear proliferation, all of which have made the nuclear option a highly contentious one in many countries. The hostility to nuclear power is in part left over from early plant designs, which both were less safety conscious and generated more waste than the models now available.

Between 1980 and 2005 the efficiency of output from U.S. nuclear power plants (the percentage of energy achieved compared to the theoretical maximum) went up from 60 to 90 percent, and the next generation of power stations will generate even less waste for each unit of energy they produce. Because of this, the waste from this new generation of power plants will add only a small fraction to the amount that already exists in countries like the United States that have relied on nuclear power in the past.

There is still the issue of how to dispose of the high-level waste, which will remain radioactive for thousands of years. But Finland, Sweden, France, the United Kingdom, and the United States are all investigating geological options for burying the waste in apparently stable rock formations.

Safety is, of course, a paramount consideration for nuclear power, though we should also point out that, historically, many other forms of power generation have come with their own hazards. Even in the developed world, coal miners are still dying in underground accidents and from silicosis—the bitter legacy of breathing so much coal dust for so long.

However, the dangers of a wider nuclear accident have led designers of nuclear power plants to make significant advances with regard to safety. Perhaps the most interesting is the new drive toward "walk-away safety," in which the machine is designed to shut itself down immediately should anything go wrong.

In the new Westinghouse AP1000 (which is currently available, though hasn't yet been used on a commercial scale), large tanks of water sit inside the reactor containment. If there is a drop in pressure in the main cooling circuit, a valve opens in the tank and the water falls down to cool the reactor. The valve is fail-safe in that it needs power to keep it closed, so if the power cuts off it also opens automatically. The upshot is that in case of trouble, the operators could walk away and the reactor would shut itself down.

Another design that uses the principle of walk-away safety is the South African Pebble Bed Nuclear Reactor, so called because its fuel is made up of "pebbles" of graphite containing thousands of tiny fuel particles. Over a period of several months, the pebbles slowly make their way to the bottom of the reactor, where they are extracted, inspected for damage, and then blown pneumatically back to the top. This effectively makes it impossible to melt the core. The first demonstration plant is due to begin operation in South Africa in 2011.

Another objection often raised is that peaceful nuclear power might encourage the dangerous proliferation of nuclear weapons. Although the materials for a nuclear bomb don't arise automatically in a nuclear power station, they can be separated out given the technology and the time. Designers are working hard to address this for future generations of nuclear power stations, but it's unlikely they will be able to eliminate the threat completely.

However, the Treaty on Non-Proliferation of Nuclear Weapons has been signed by nearly 190 countries, and the thorough inspection routines by its watchdog committee, the International Atomic Energy Authority (IAEA), have to date picked up suspicious activities extremely quickly. For instance, they successfully blew the whistle on countries such as North Korea and Iran—which were using their nuclear power plants to mask their attempts to make weapons—thus focusing the world's attention on the problem. Moreover, the motivation for any country to try to develop nuclear weapons has very different origins from the need for energy from nuclear power plants. North Korea did not seek to develop weapons because France had nuclear energy.

Nuclear power currently accounts for about 5 percent of global energy, and in some quarters it is predicted that this could double by 2030. Because of the advances that have been made in addressing the apparent disadvantages of nuclear power, and because it is one of the very few low-carbon technologies that is ready to go, we believe that a further generation of nuclear power stations will be at least a necessary stopgap in countries that already possess the technology, and perhaps in others, too, while we wait for other low-carbon energy sources to become available.

If your gut instinct is to react against nuclear power, consider the arguments here very carefully. Individual and community support will be vital for new nuclear power stations to be developed, and the problem of climate change will require us all to make some very hard choices. Nuclear power is one of the very few low-carbon technologies that are already on hand, and though it is not necessarily an ideal way to make energy, the dangers of climate change are certainly far worse.

Wind Energy

Wind sounds like the perfect solution to the energy crisis. The air moves around anyway, so why not harness its energy for free, with no carbon emissions to boot? Partly because of this, wind turbines are springing up around the world. Installations have been growing by 28 percent a year since 2000, with a record 40 percent leap in 2005. The United States has an impressive thirteen gigawatts of wind capacity, spread across thirty-two states.[12] (Texas has the highest capacity for an individual state,

followed, unsurprisingly, by California.) One of the most visible
wind farms in the United States is also the only one located on
the coast. Its five giant wind turbines sit on the road to Atlantic
City, New Jersey, and are passed by thirty-five million visitors to
the city every year.[13]

However, wind still makes up a mere two-tenths of a percent
of the global energy mix. One reason is that wind energy comes
with its own public perception problems. There are claims that
turbines are noisy, and that they cause problems for radar and
airline flight paths, as well as for unwary birds and bats that
stray into their blades. But most of all, many of the best places
to put them on land are big, windy, open expanses, which are
often also very beautiful. Even supporters of renewable energy
rarely want to see giant turbines marching across their precious
wildernesses. Much of the hard work in setting up the Atlantic City wind farm involved persuading the local community to
come on board. And in the United Kingdom, while just over two
gigawatts of potential wind energy have made it to the national
grid, another nine gigawatts are currently mired in contested
planning permissions.

As with nuclear power, this is an area where your individual attitude can contribute to a big collective difference. If a
wind farm is proposed for land near where you live, it's worth
thinking about your reaction very carefully. It's not necessarily
a good idea to put turbines up in places that are outstandingly
beautiful and wild. But those aren't the only sites where wind
power could be productive. Ultimately, wind will have to be a
large part of the overall mix for tackling climate change, and
none of us can afford to dismiss it out of hand from our own
backyards.

The more successful onshore wind-farm schemes have involved starting with small-scale installations and working up.
It turns out that turbines are neither noisy nor particularly ugly

(especially compared to electricity pylons), and when one or two are in place it's easier to be enthusiastic about scaling up. Some small, windy towns have discovered that a large wind turbine can provide electricity to the whole town at very little cost. In these cases, far from being an eyesore, the turbine can become a testimony to the smart—and responsible—action of the townspeople. If you live in such a town, it's well worth talking to your local representatives about this.

Another possibility is to place wind farms offshore, where there would be fewer worries about blighting the view. However, the windiest places also tend to be the ones with the harshest conditions. New materials will need to be developed to stand up to the buffeting they would receive, and transportation of the generated electricity could also be a problem. Thus, this approach is in its infancy even in countries that have very windy coasts.

One final potential problem with wind is that, like many other renewable sources, it is inherently fickle. Denmark, which has five gigawatts of installed wind power, has managed this by exporting electricity to its neighbors Norway, Sweden, and Germany when the wind blows, and importing—for instance, from Norway's hydro plants—during lulls. Still, for those countries that do not have alternative providers so near at hand, the sporadic nature of wind will remain a problem, at least until we can come up with ways of storing wind electricity along with better methods of forecasting when the wind is on its way.

However, none of these problems is insurmountable, and wind will certainly have to be a vital component in the future low-carbon energy mix, with the most promising sites being in northern Europe, on the southern tip of South America, in Tasmania, and in North America in the Great Lakes and along the northeastern and western coasts. One study predicts that, in principle at least, wind could generate 126,000 terawatt-hours of electricity per year by 2030. That, by the way, would be five times the global electricity requirements.

Solar Heating

Enough sunlight falls on Earth to meet our energy needs ten thousand times over. However, it's highly dispersed. Various technologies already exist to concentrate the sunlight using parabolic mirrors and focus it on tanks of liquid—usually water or oil. If you're lucky enough to live in a sunny part of the country, solar heating panels like these are an excellent way to heat water for individual homes. They sit on the roof rather like a satellite dish and provide baths and showers that are satisfyingly hot and guilt-free.

Solar heating can also be used on a much larger scale, by making the hot liquid drive an engine to generate electricity. Clouds can get in the way, so electricity from solar heating would have to be supplemented from time to time by another source (probably coal, natural gas, or bioenergy). The best places to site solar-heating power stations would be in low-latitude deserts, which receive intense sunlight and suffer relatively few cloudy days. The American Southwest, particularly the Mojave Desert, has an excellent combination of high levels of sunlight and low levels of cloud. Solar-heating plants have been installed there since the late 1980s, with new projects under construction and others currently being planned.

New projects are also either being built or planned in eleven other countries, including Australia, Spain, Israel, and Morocco. Analysts estimate that by 2040, solar heating could be meeting 5 percent of the world's electricity needs. However, the potential is much greater. If only 1 percent of the world's desert areas could be linked to the rest of the world by high-voltage cables, that would be enough to meet the world's entire electricity needs, as forecast for 2030.

Solar Photovoltaic (Solar Panels)

Unlike solar heating, this technology uses individual rays of sunlight to dislodge electrons inside an electronic panel. It thus generates an electric current directly instead of having to go through the intermediate step of using hot liquid and an engine.

Estimates vary, but there are enough solar panels like this around the world to generate at least five gigawatts (which is just 0.004 percent of total world power). Solar panels have the major advantage that they can be placed directly where the energy is required, which avoids the need for power lines. Thus, they are especially popular in rural areas of developing countries, where normal grid electricity is either unavailable or unreliable. However, the panels remain too inefficient at converting sunlight to electricity, and too expensive to manufacture, to make them fully competitive with other forms of electricity in the developed world.

One reason they are expensive is that solar panels tend to be based on silicon, because industry already has plenty of experience making silicon chips for computers. A cheaper alternative would be to use plastics, ceramics, or some other material, perhaps devised through the emerging science of nanotechnology. This will require technological breakthroughs and probably won't be available for several decades, but it does have fantastic potential to yield an energy source that is cheap, free of emissions and waste, and has an unlimited fuel supply (the sun). Solar panels could also be used in conjunction with direct solar heating so that you could even exploit the heat generated by the electronic chips themselves.

Carbon Capture and Storage (CCS)

This could be the most important bridging technology between using fossil fuels and new, low-carbon alternatives. It's so important, in fact, that many people are now talking about CCS as if it were already established, though as yet it is only at the pilot stage.

The idea is to burn fossil fuels in the normal way to generate power, but then to grab the carbon dioxide and bury it before it can escape into the atmosphere. This has the great advantage that it can remove emissions from traditional fossil-fuel plants, thus buying the world some time to develop new low-carbon alternatives. CCS is likely to be especially important for countries like India and China, which are currently exploiting their vast coal reserves at an increasing rate to fuel extremely rapid economic growth.

Critics of CCS say that it would allow us to keep our old, bad polluting habits and discourage fossil-fuel companies from switching to alternative forms of energy. But it's now too late to raise this sort of objection. The typical lifetime of a coal-fired power station is forty years, and developing countries with large coal reserves are bound to make use of them. Even with strong action to encourage the use of renewable and low-carbon energies, the International Energy Agency predicts that fossil fuels could still make up half the energy supply in 2050.[14] If we are to have any chance of reining in climate change, some form of CCS will be a vital part of the mix.

The technology is likely to work best when applied to a single large source of carbon dioxide, such as a power plant or the chimneys from a major energy-intensive industry. Studies suggest that CCS could reduce emissions from fossil-fuel power plants by 80 to 90 percent, though this will come at a price. The present technologies for capturing carbon are still expensive, and electricity generated from a fossil-fuel plant using carbon capture will always cost more than one that vents its carbon dioxide into the atmosphere. For CCS to be economically viable there will have to be some kind of pricing system that rewards reductions in carbon emissions.

In principle CCS can be applied to any design of power plant, though in practice it is much cheaper and more efficient when used in modern "cleaner" designs that already produce relatively concentrated streams of carbon dioxide in their outflow. (These designs also produce much less local pollution, although they tend to be more expensive and are rarely used in the new power plants springing up in the developing world.) Also, and most important, scenario studies suggest that, regardless of the design of power plants, CCS will always be both more efficient and cheaper to implement if it is built into the design of the plant rather than retrofitted at a later date. Given the rate at which new power plants are now being built, especially in India and China, it is imperative that CCS systems are installed in pilot schemes in a range of different power-plant designs as soon as possible. Studies suggest that, technologically at least, CCS pilots could be up and running within five to ten years, but that will require a combination of political will and economic incentive.

Once captured, the carbon dioxide can be compressed into liquid form to be transported via pipeline or ship, or "fixed" into some inert solid chemical compound. As for storage, the likeliest option would be to inject the gas into some deep geological network of tunnels and cracks, such as an empty oil or gas field or a salty aquifer. (Freshwater aquifers tend to be in use already to provide water for drinking and irrigation.) Empty oil and gas fields would probably be the easier option, since they exist in large numbers and we already know that they have been capable of holding on to their contents for hundreds of millions of years. However, coal-rich countries such as China and India have no access to empty oil fields of their own. Thus, the challenge will be to find a secure way of exploiting aquifers. There should, however, be plenty of space available. The IPCC estimates that known gas and oil fields could store between 675 and 900 gigatons of CO_2, and that saline formations could take at least another 1,000 gigatons,[15] while other reports suggest that this is a very conservative estimate.

In the United States, seven Carbon Sequestration Regional Partnerships have been set up to investigate the possibilities for CCS. These are regional-scale public–private partnerships that include state agencies, universities, and private companies, and span forty-one states, two Indian nations, and four Canadian provinces. They are currently in the process of conducting small-scale field tests for carbon storage, but the next phase of the project should involve larger-scale attempts to integrate capture and storage.[16] The results should also feed into FutureGen, a $1.5 billion project involving public and private partners to design, build, and operate a coal-fueled, near-zero emissions power plant, at an estimated project cost of $1.5 billion. However, as yet this is only at the site selection phase.[17]

In 2007 the UK government announced a competition for funds to develop a CCS demonstration plant. The European Union is funding a similar demonstration plant to be sited in China, under the Near-Zero-Emissions Coal Initiative.[18] The United Kingdom and China signed an agreement in 2005 to detail specifically UK-funded action on the project, and a similar agreement was signed between the European Union and China the following year, with the aim of having a demonstration project up and running by 2014. Projects like these should help to identify and overcome technical difficulties associated with capture and storage.

Opinions differ over the likely spread and scope of CCS in the future. Some scenarios suggest that CCS will never be more than a bridging technology that will be deployed rapidly but then superseded in the middle of the century by other low-carbon technologies. Others assume that it will be widespread until at least the end of the century.[19] An IPCC special report on CCS, published in 2005, calculates that between 30 and 60 percent of global emissions of CO_2 from power generation could be suitable for capture and storage by 2050, and that a further 30 to 40 percent of emissions from industries such as cement making could also be amenable to capture by then.[20]

An enticing prospect is that CCS could be used to capture and store the carbon dioxide from burning biomass. In that case, the result would be a net loss of carbon dioxide from the air, which could help with achieving that low stabilization target of 450 ppm CO_2eq.

Nuclear Fusion

As mentioned above, "normal" nuclear power works by splitting an atom. By contrast, fusion works by fusing the nuclei of two atoms together. This is the process that fuels the sun, and it has several distinct advantages. For one thing, it produces virtually no radioactive waste.[21] For another, the feedstock is very plentiful. Nuclear fusion uses a naturally occurring, heavy form of hydrogen that is abundantly present in seawater, along with lithium metal, which is abundant on Earth. (It is used, for instance, in laptop computer batteries.) The waste product, nonradioactive helium gas, simply floats away.

The problem lies in persuading those two nuclei to overcome their natural distaste for each other long enough to merge and release copious amounts of energy. Copying the sun's strategy means heating the gas to extraordinarily high temperatures. That takes a lot of energy in the first place, and also requires materials that can survive the heat.

The Joint European Torus (JET) in the United Kingdom has produced sixteen megawatts of fusion power by confining gas in a magnetic bottle and heating it to more than 100 million degrees, and the Japanese torus at Naka, which is modeled on the JET, has achieved similar results. These two projects are the largest fusion devices in the world. However, a fusion power station will need to be some ten times the size of these projects and operate around the clock.

Acting on behalf of the European Union one of us (David King) helped to broker an international consortium to address this problem. Comprising the European Union, Russia, Japan, China, South Korea, India, and the United States, this is the biggest ever international technological enterprise. Uniquely, it spans countries from the developed and developing worlds, from both East and West. Each country is contributing a hefty €500 million, and the European Union and France even more.

Together, the consortium is about to begin constructing an experimental device, called the International Thermonuclear Experimental Reactor, in France. The device will cost some $10 billion, will be on the scale of an eventual power station, and will operate for twenty years in an effort to resolve the scientific and engineering challenges of fusion. If all goes well, the first commercial fusion power could be available within the next forty years—or sooner, if the governments involved increase their investment in the project.

Final Note

Most of the technologies and approaches that we have described here are ready to go, or will be soon. The rest are ripe for investment. The International Energy Agency predicts that some $20 trillion will be invested in energy over the next twenty-five years.[22] If a significant part of this goes into fossil fuels, rather than the low-carbon alternatives, it will be very difficult to go back. Even technologies such as carbon capture and storage, which work alongside traditional fossil-fuel power plants, would be much more expensive to retrofit than to design into new power stations from the start. In other words, the technological picture tells the same story as the science: The need to change our carbon habits is urgent, and the time to start is now.

PART III

POLITICAL SOLUTIONS

By now it should be clear that the world needs to reduce its greenhouse emissions and has both the technical wherewithal and ingenuity to do it. But the science and the technology are the easy part. They provide the way. What we need next is the will. This section of the book covers the economic and political aspects of climate change. Much of this is necessarily the concern of big business and nations, but individuals are still important as a source of pressure from below. These chapters will explain what to look out for in political agreements. They are a handbook, if you like, to show you whether your politicians are doing what they should. At the end of the section, we put all of these solutions—both technological and political—together to show how your personal choices can make a difference.

10

IT'S THE ECONOMY, STUPID

Pay Now or Pay Later

For a while now, economics has been at odds with science on the issue of climate change. Almost from the beginning, science has been warning us to act now before it's too late. But until recently, some economists have instead been telling us to wait until we're all richer, or until new technologies come along.

That's because many economic theorists have consistently argued that kicking the carbon habit will damage the world economy, producing serious economic pain for relatively little climate gain. Even some of those who accepted that climate change was more than a soppy green theory have tended to argue that it would be better to wait until economies had grown and the world was richer; that way we could use our future wealth to tackle what was largely perceived as a future problem.

This argument was apparently confounded in 2007 when UK economist Sir Nicholas Stern led a team that published the world's most extensive review, the *Stern Review on the Economics of Climate Change,* commissioned by the British government. The *Stern Review's* six-hundred-plus pages contained many detailed analyses of the economic blessings and curses of climate change. But the one that made headlines around the world was this: According to Stern, tackling climate change immediately to the tune of a maximum value of 500 to 550 ppm CO_2eq would cost just

1 percent of global per-capita consumption per year; leaving the problem to deal with later would raise the bill inexorably; and letting climate change continue unabated could eventually cost the world a full 20 percent of global consumption per year.[1]

Paying 1 percent now to save 20 percent later sounds like a good deal in anyone's books. But other economists immediately leaped on Stern's assumptions, claiming that he had built this answer into his models. More conventional models, they said, would give a much lower future benefit, and probably a higher present cost. Stern vigorously defended his choices, and the battle rumbles on.

Why can't these economists agree? The reason illustrates how economic models differ fundamentally from scientific ones, and also—we would argue—why in the end they can't tell us with any degree of certainty how much it will cost us in the future if we ignore climate change today.

The economists' argument is a bit technical, but it is worth examining. Because once you understand the origin of the Stern spat, you will also understand just how little faith you should have in any bold, specific economic assertions about the future costs of climate change.

Generally speaking, economic models that are looking at the effect of any change try to work out whether it will make people in the future better or worse off. To put it at its simplest, suppose something happens that makes me $5 richer but costs four other people $5 each. On balance, the five of us are worse off.

Models looking at the consequences of climate change do something similar. Though the models are more complex and include many more factors, it's still more or less a case of adding up the consequences for everyone in the world to get a figure for global well-being that might be positive (on the whole, the world will be better off) or negative (on balance, things will get worse).

There is, however, an extra twist, which economists call discounting. Discounting the future is economistspeak for the "bird in the hand" principle. The idea is that a loaf of bread is worth more to you today than it would be next week because the future is uncertain. Looking on the bright side, by next week you might have won the lottery and be rich enough not to care about the price of bread. Looking on the dark side, you might no longer need it because you've been run over by a bus. Either way, it counts for more now than it would then, and the difference is called the discount rate.

The discount rate that Stern chose for his climate calculations was unusually low. In other words, a loaf of bread would be worth almost as much to you in a week's time as it would in your hand right now. This makes problems in Stern's future cost much more than many economists are comfortable with. By choosing a low value for the discount rate, claimed his critics, Stern was skewing the costs of future climate change so they counted for more than they should.[2]

Stern points out that since the time frame for the global warming problem runs over many generations, discounting heavily is equivalent to saying that people born in the future matter less than we do today. In the above example, it would be like saying a loaf of bread is worth more to you than it would be to your unborn grandson. Making this kind of assumption about future generations who aren't here to stand up for themselves is something that Stern finds unethical.

He also makes the undeniable point that if we can say "let's leave the problem for the next generation," the same argument could hold for them, too. By taking this route, humans would consistently pass on the challenge of dealing with climate change to successively more remote descendants, while the world crashed and burned around them.[3]

Nonetheless, by his choice of discount, Stern does seem to have encouraged the future to cost as much as possible.

Economist Sir Partha Dasgupta[4] from Cambridge University points out that we don't actually know how big a difference these choices made to Stern's findings. The trouble, he says, is that the future cost of excessive carbon is extraordinarily sensitive to the discounting figures you choose. The IPCC report agrees. It says that discounting the future at 3 percent means that one ton of carbon bears an ultimate "social cost" of about $62. Dropping the rate to 2 percent increases the cost to $165 per ton. Set the discount rate at 0 percent and the number goes up to a massive $1,610 per ton.[5] That's a big difference in cost for something—how important we rate our grandchildren—that is an ethical rather than a financial call.

The bottom line is that the hard numbers of the future economic costs of climate change depend very sensitively on what are essentially touchy-feely choices. Put that way, any attempt to put a firm price on future damage seems hopeless.

This problem is exacerbated when you add in the risk that Earth's climate could tip over one of the thresholds that we described in chapter 5. The level of uncertainty that this brings to economic models makes concrete predictions all but impossible. As Dasgupta puts it: "Climate change and biodiversity losses are two phenomena that are probably not amenable to formal, quantitative economic analysis. We economists should have not pressed for what I believe is misplaced concreteness. Certainly, we should not do so now."[6]

However, even if we can't know the exact economic costs of climate change in the future, we still know a changing climate will bring direct misery—flood, drought, and famine—to very large numbers of human beings, along with all the indirect dangers associated with mass migration and resource wars. In fact, it's already doing so. You could argue that there's little need to worry about how much climate damage might cost us in the future when its effects are already being felt today. And many of the future variables are in any case incalculable. What value would we place on leaving Venice intact for future generations? Or London? Or Cairo?

Bearing in mind all these problems, how do we decide whether it's a good idea to act now or leave the problem for later? Perhaps the best way is to ignore the "misplaced concreteness" of the economic predictions and listen instead to the science. As we explained in chapter 6, if we want to avoid leaving future generations with very serious—even dangerous—changes in climate, we need to stabilize greenhouse gas levels at 450 ppm CO_2eq. And to do that we have to act now. To get greenhouse gases down to this level, we need global emissions to peak within the next fifteen years. If we waited to act, 450 ppm would very soon be beyond us. (In fact, even higher stabilization levels would require us to act in the next twenty years.) The coming decade is the only window of opportunity we, or any other generation, will ever have.

Put this way, the economic question becomes not so much "How much will it cost us if we don't pay to stop it now?" but "Can we afford to pay if we do?" And the answer is that the cost of reducing emissions could be surprisingly low.

How Much Will It Cost?

The *Stern Review* considered a wide range of potential costs that could be incurred in the tackling of climate change, as well as savings from improving efficiency. Overall, it concluded that the annual cost of putting ourselves on a path to stabilizing the atmospheric levels at 550 ppm equivalent or below by 2050 would be between −1.0 and +3.5 percent of GDP, with an average of around 1 percent.[7]

A report in the business journal the *McKinsey Quarterly* has an intriguing way of considering the problem.[8] The authors made a list of all possible ways of reducing greenhouse gas

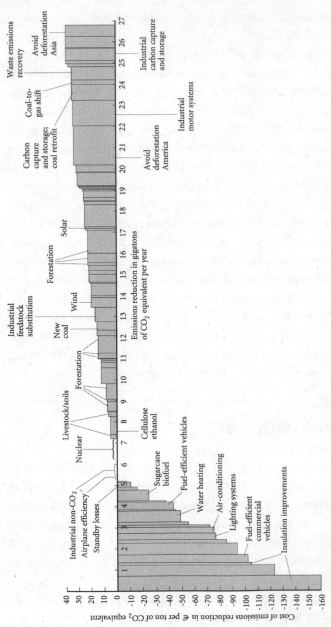

The cost of reducing greenhouse gases using different technological approaches up until 2013. Anything below the axis saves money, while anything above the axis costs. All the strategies shown here, taken together, would be enough to keep greenhouse gases below 450 parts per million. The balance of costs and savings also shows that the overall cost of this would be relatively low. (Source: "A cost curve for greenhouse gas reduction," P. Enkvist et al., *The McKinsey Quarterly*, no. 1, pp. 35–45, 2007)

emissions up to a "marginal cost" of €40 (almost $60) per ton. (That's to say that many of their methods actually save money; many cost a little, but still much less than €40 per ton; and none costs more.) Then they line these different ways up on a graph.

The graph bears all the usual suspects. It starts on the far left with the methods that will bring the biggest savings: building insulation (very big savings of more than €150, or roughly $220, per ton); improving efficiency in lighting (savings of €100, close to $150, per ton); heating (savings of €50, or nearly $75, per ton); and so on. At about zero net cost is reducing industrial emissions of greenhouse gases other than carbon dioxide. Then the price begins to go positive for nuclear energy (about €1, or $1.50, per ton), up to carbon capture and storage on coal power plants (net cost of €30, or about $45, per ton), on up to the most expensive methods, such as waste emissions recovery and industrial carbon capture and storage (just short of €40, $60 or so, per ton).

Going that far, say the authors, would produce enough overall savings between now and 2030 to set the world on a stabilization path for greenhouse emissions at 450 ppm equivalent. Encouragingly, their approach uses only methods that either exist now or are in the immediate pipeline (such as carbon capture and storage). It doesn't require any immediate technological breakthrough.

The authors also add up all their savings and costs to arrive at an overall price for following this path. They say that using every method they describe to its fullest potential would mean that by 2030 we would be paying out only about 0.6 percent GDP. Even if some of the methods fell short, it would still cost just 1.4 percent GDP by 2030.

Could we afford to pay these sorts of sums? That's more a political than an economic question. Dasgupta points out that 1 percent GDP globally would mean a cost of about 1.8 percent GDP for the rich countries who would need to bear the brunt of the costs, a sum that adds up to more than seven times the annual global aid budget. Put that way, the task of persuading our governments to pay out sounds hard. It becomes tempting to think of the money needed to pay for climate change as a sort of opportunity cost. Why don't we just spend the money directly on aid instead, and save lives right now instead of waiting till later?

However, that's a false argument. We don't have a single pot of money that the world has contributed to, and that we now need to decide how to spend. If the money doesn't go toward dealing with climate change, it's very unlikely to go instead to aid. Moreover, in many cases the money to deal with climate change is already there in a different pot, marked "energy." As we mentioned in chapter 9, the International Energy Agency predicts that more than $20 trillion will need to be invested between now and 2030 to meet the world's growing hunger for power. The question we face regarding this pot is whether it should be invested in fossil fuels or in low-carbon sources.

Also, we believe that the model of thinking about the cost of climate change as an alternative form of aid is misguided. This isn't about handing over money to poor people to make their lives a little better. It's about changing the culture of the entire planet so that we leave behind a climate that all our grandchildren will be able to weather. The next two decades are our only possible window of opportunity to rein in greenhouse gases to a level that will achieve this goal. After that, however rich some parts of the world have become, it will be too late.

Because of this, we believe it's better to think of the costs of climate change not so much as a handout as an insurance policy against a dangerous future. The authors of the *McKinsey Quarterly* article point out that even excluding life insurance, in 2005 the global insurance industry's turnover was 3.3 percent of GDP, which makes the cost of dealing with climate change seem much more reasonable. In fact, investing in new forms of energy and claiming all the savings from improving our efficiency could even boost our economic growth.

How Do We Pay?

Normally, we rely on market signals—such as prices—to make decisions about how to spend our money. But the price we currently pay for energy takes no account of the effect it has on the climate; it gives no encouragement to do things the low-carbon way, nor does it provide economic penalties for taking a high-carbon path. Because of this, the *Stern Review* describes climate change as "the greatest market failure the world has ever seen."[9]

To rectify this, many economists advocate that we adjust the market. The simplest way would be to put a tax on every ton of greenhouse gas emitted. That way, the price of any activity that created greenhouse gases would reflect its true cost to the environment, and the market would clamp down on the biggest offenders, making it cost-effective for them to turn to low-carbon solutions.

There is some evidence that this can work, at least on a national level. In the early 1990s Norway introduced a carbon tax on emissions from energy, and it did seem to encourage technological innovation.[10] However, the problem comes when you try to apply this approach globally. Agreeing on international taxes is notoriously hard—the European Union has experienced terrible difficulties in trying to regularize the complex and myriad differences in its members' sales taxes. And although Sweden, Finland, and Denmark also introduced carbon taxes during the 1990s, they have not harmonized their approaches with Norway or with each other. If such similar countries can't come to agreement, there is little hope for doing so with the vastly more disparate countries in the rest of the world. Moreover, it's hard to imagine any country welcoming, or even tolerating, an annual tax bill from the World Bank.

An approach more likely to work politically is the "cap-and-trade scheme." This is modeled on the successful approach used in the United States in the mid-1990s to reduce sulfur dioxide emissions and thereby cut down on acid rain. The "cap" part is that every participant in the scheme has a limit on how much it can emit in a given year. The participants are then awarded a set of permits equal to that cap. The "trade" part is that anyone going over that limit can buy permits from anyone who has emissions to spare. Every year the cap gets a little lower, and emissions fall inexorably—aided by the invisible hand of capitalism.

One clever thing about cap and trade is that it ensures that the fall in emissions occurs as economically as possible, since if it's cheaper for me to pay you to cut emissions than to cut them myself, I can buy your permits. This worked beautifully in America for reducing sulfur, as emissions dropped faster than most people had imagined, and at a fraction of the cost that doomsayers had been predicting. In 1999 BP set up an internal market for emissions trading of carbon dioxide among their various subsidiaries. It cost $20 million to set up, and by encouraging energy efficiency over the following three years saved the company an impressive $650 million.

The United Kingdom was the first country in the world to set up a national carbon-emissions-trading scheme, in 2004, and a Europe-wide scheme swiftly followed the next January. This was the world's first international trading scheme for carbon.[11] It's open only to members of the European Union, but—especially with the addition of ten countries from Eastern Europe in 2004, and a further two in 2007—the EU countries are now so varied, and differ so greatly from one another in economic development, that they make an effective microcosm showing how a worldwide trading scheme might look.

All the energy generators, metal manufacturers, and producers of cement, bricks, pulp, and paper in each member state are obliged to adhere to an emissions quota. (Together these industries account for almost 40 percent of the European Union's overall emissions.)

The process sets a natural price on carbon emissions per ton that arises out of the standard economic model of supply and demand. In the European scheme, the price per ton was holding up well at between €10 and €20 (roughly $15 and $30) for the first year of trading, a value that would certainly make emissions savings worthwhile. However, in April 2006 disaster struck. Actual emissions figures from the industries involved during 2005 were much lower than the ones that had been predicted. The price of carbon plummeted, and permits for the first phase became almost worthless.[12]

This was a painful lesson in the hazards of setting up a brand-new market. It's now clearer than ever that for a market like this to work, it must create real scarcity in permits, making the right to emit greenhouse gases a rare and valued commodity. The European Union has adapted the project to meet this requirement. The next phase, which runs from 2008 to 2012, has now set much more stringent caps, and advanced permits for that phase are already maintaining their prices well at between €15 and €20 per ton.[13] So, in spite of these teething troubles, the market is now well established, and it will surely be a powerful tool for reducing emissions in a region that adds up to the world's largest combined economy.[14]

Similar schemes have been springing up in the United States. California now has its own carbon-trading scheme, as does a consortium of northeastern states, and the Chicago Climate Exchange has been operating successfully (dealing with voluntary emissions caps) since 2003. Elsewhere in the world, in New South Wales in Australia, emissions trading has been mandatory for electricity companies since 2003, and the prime minister recently announced that there would be a nationwide scheme in place by 2012. When the Kyoto Protocol came into effect in February 2005, it, too, involved an emissions-trading scheme for those countries that had ratified the international climate treaty.

In May 2007 the World Bank reported that the global carbon market had tripled in the previous year, growing from $10 billion in 2005 to $30 billion in 2006, most of which came from the sale and resale of European Union Allowances,[15] which is certainly encouraging. Meanwhile, as well as tightening up their emissions, the European Union is investigating ways to bring new sectors under the trading umbrella, especially transportation and international aviation and shipping, the latter two particularly vexatious since they are currently not even counted in an individual country's emissions.

Carbon markets won't, in themselves, be enough. The 2007 World Bank report on the state of the carbon market noted that "the enormity of the climate challenge . . . will require a profound transformation, including in those sectors that cap and trade cannot easily reach."

For instance, agriculture, private transportation, and homes would involve far too many individuals to be amenable to the sort of monitoring and auditing that a cap-and-trade scheme would require. In cases like these, governments will also need to oblige energy companies to use more renewable energy, to introduce standards for efficiency, and, above all, to point everyone in the right direction by providing reliable information about how much greenhouse gas an individual product is responsible for producing throughout its lifetime.

The World Bank report also singled out the need for individual governments and international consortia to encourage investment in developing new, low-carbon technologies. One important way to do this will be to set up public–private partnerships to encourage investment in research and development, such as the United Kingdom's Energy Technologies Institute.[16]

The Robin Hood Effect

Most analysts agree that the burden of payment should fall most heavily on the world's richest countries, which are not only in the best position to pay but have also been, historically, the main cause of the greenhouse problem. But the costs will not fall evenly. In another of the nasty ironies associated with the greenhouse problem, solutions that actually save money (such as insulating homes) tend to fall on the rich nations, whereas the solutions that cost the most (such as capturing carbon from Indian power plants or preventing deforestation in Brazil) fall on the poorer nations, who have done the least to create the problem in the first place. Any arrangement for who pays, and how, will have to ensure that money from the richer nations passes effectively to the poorer ones, specifically targeted toward cutting emissions in the most economically efficient way possible.

In principle, cap and trade allows funds to be transferred between nations. But in practice, we need to go farther to ensure that money flows from rich countries to poorer ones. This isn't just some twenty-first-century Robin Hood principle of generosity to the poor. In fact, it is the only way to encourage developing countries to leapfrog the old, bad polluting habits of the industrial world and invest directly in low-carbon habits instead—which is something the entire world needs them to do.

The Kyoto Protocol brought in two mechanisms to channel money from the most industrialized countries to those that are less developed. In the jargon of treaties, these are known as Joint Implementation (JI) and the Clean Development Mechanism (CDM).

Joint Implementation means that one developed country can gain carbon credit by setting up a program to reduce emissions in another relatively well-developed country. For instance, France could pay to set up efficiency improvements in Romania. (Joint Implementation works mainly for countries from Eastern Europe and the former Soviet Union, which are relatively developed, but where a small amount of investment can make a big difference to energy efficiency.)

The Clean Development Mechanism is similar, but takes place between developed and much poorer developing countries. Thus, the United States could gain credit by setting up a program to reduce emissions in India. The idea is that this brings two benefits: Emissions reductions become relatively cheap for the developed country, while developing countries receive funds that enable them to adopt new, low-carbon technologies.

One of the biggest potential problems with this approach is ensuring that the projects included really do reduce overall emissions, and that they wouldn't have happened anyway. The Europeans are already learning important lessons in this regard from the use of the CDM in the European Emissions Trading Scheme. In June 2007 the World Wildlife Fund (WWF) published a report called "Emission Impossible," in which it highlighted problems with the way the European scheme uses the CDM.[17] Developing countries that have ratified the Kyoto Protocol can put forward proposals for projects that they say will reduce carbon emissions. These are then assessed by a CDM Executive Board based in Switzerland. The idea is to check especially whether the projects will deliver the proposed emissions savings, and whether the savings are "additional," in that they wouldn't have happened without the extra funding provided through the CDM. Moreover, the board is supposed to check that the projects fit with the general aims of sustainable development in the countries involved.

The WWF has come up with a list of approved projects that, according to the organization, don't meet one or more of these criteria. For instance, it reports that a hydropower project in China was accepted in 2006 even though the CDM board received a submission saying that "revenue from CDM credits was irrelevant to the decision to go ahead with the project," and adding that "construction began in October 2003." An article in *Nature* also criticized the CDM,[18] and in Britain the *Guardian* claimed that up to 20 percent of all the carbon credits issued so far under the CDM could be doubtful.[19]

The solution to this is simple, as long as it's effectively implemented. The WWF recommends that, in the future, CDM projects should be certified by the "Gold Standard."[20] This is an independent benchmark set up to accredit schemes to cut emissions in developing countries for both international and individual investors. The Gold Standard, for instance, admits only renewable energy and efficiency projects and is highly conservative in its estimate about whether a project is truly "additional."

The Gold Standard would certainly provide a much more reliable guarantee that the required emissions will be achieved. On the other hand, by excluding projects such as reforestation, it is also somewhat limiting. Better still would be a new standard, agreed to by governments, that would take into account the full range of possible ways to reduce emissions in the developing world and provide a reliable method of emissions accounting even for the projects that are harder to quantify.

All these problems notwithstanding, the CDM is clearly working in its endeavor to draw foreign investment into developing countries that enables them to cut emissions. The World Bank report on 2007's carbon market says that the Kyoto Protocol mechanisms have generated $8 billion in new resources for developing countries, which have in themselves attracted a further $16 billion in associated investments for generating clean energy. When these initial problems with the CDM have been resolved, it will be a vital component of the climate tool kit.

A final note on this topic concerns the question of deforestation. The *Stern Review* identified halting the rate of deforestation as one of the most cost-effective means of rapidly cutting carbon dioxide emissions. In principle, it's obvious that slowing the rate at which forests are cut and burned will reduce carbon dioxide emissions. However, the Kyoto Protocol, and thus the CDM, found this too hard to turn into a formula, and therefore stuck to reforestation and afforestation projects.

The *Stern Review* notes that a hectare (about 2.5 acres) of forest generates an income of just $2 per year if converted to pasture, $1,000 per year for soybean or palm oil production, or between $240 and $1,035 as a onetime return for selling timber. It then calculates that, even assuming a modest carbon price of $30 per ton, the same hectare of forest could be worth $17,500 if kept intact.[21]

At present, though, there is no provision for earning credit by protecting forests. This leads to some obvious absurdities. For instance, environmental groups claim that forests in Malaysia and Indonesia are being cut down and burned to clear land that is then used to grow sugarcane for "low-carbon" biofuel.

After pressure from scientists, in 2007 the United Nations Framework Convention on Climate Change launched a two-year consultation for a mechanism to encourage the reduction of deforestation levels, provisionally called "Reducing Emissions from Deforestation" (RED).[22] One approach, proposed by Papua New Guinea and Costa Rica on behalf of the Coalition of Rainforest Nations, was to mimic the cap-and-trading schemes with a separate scheme built around deforestation credits. The idea is that each nation establishes a baseline rate of deforestation, converted into tons of carbon dioxide. Any reductions below this baseline could then be sold in a carbon market. Emissions above the baseline would automatically disqualify a country from trading.[23] This would also require some solution to the problems that Kyoto Protocol negotiators encountered: how to estimate a reliable baseline for the rate of deforestation before intervention; how to ensure that the forest remains intact permanently rather than just for a few years, until attention passes elsewhere; and how to avoid "leakage," whereby people displaced from cutting down one part of the forest simply move elsewhere.

Adaptation

As we explained in chapter 4, every country will need to adapt to the climate change that is already occurring, and that is inevitable. The countries that will be hardest hit will be those that can least afford to pay (and that have contributed least historically to the climate problem).

There are already three global funds aimed at aiding the least developed countries to adapt. The Least Developed Country Fund takes contributions from developed nations and uses them to help the poorest countries draw up adaptation plans. By April 2006 it had received a total of $89 million in pledges and actual funds. The Special Climate Change Fund supports direct adaptation activities and has received some $45 million to date. Finally, the Adaptation Fund arises directly from the Clean Development Mechanism. Each CDM transaction pays a 2 percent levy into this fund, which is also used to support adaptation projects in the poorest countries. The World Bank estimates that by 2012 this will have raised up to $500 million.[24]

However, the first two funds are dependent on aid, which some developing countries consider to be an inappropriate model. Indeed, a levy is a more appropriate tool than relying on a country's generosity—and potentially diverting resources from other aid projects. The third fund is based on a levy, but carries the economic problem that, since it is effectively a tax on the money invested, it provides a perverse financial disincentive to invest in the CDM. (One way around this would be to apply the same levy on all carbon trades, so as to remove the distinction.)

Various alternative proposals have been put forward.[25] One attractive possibility is that a levy could be applied to some voluntary and polluting enterprise such as air travel. (This would help compensate for the longtime exemptions that air travel has had from tax on fuel duty, as well as help control the projected massive growth in air travel.) Another approach would be to come up with a separate fund levied on industrialized nations as a function of their GDP, or, more appropriately still, as a function of their historical greenhouse emissions.

The Color of Money

All in all, the problem of climate change is beginning to shift from a scientific and technological issue to an economic one. Around the world, brokers have noticed the amount of money changing hands over climate change and are eager to join in. The business world, too, has spotted that climate change comes with opportunities, as well as risks.

In some cases, companies seek to attract consumer support by taking a climate-friendly stand. For instance, companies as widely diverse as the liberal-leaning Google and the right-wing and normally antienvironmental Fox News network have recently pledged to go carbon neutral. Others seek to be ahead of the game by improving their own efficiencies, while yet others are looking to develop environmental products to sell. BP (which has dropped its original name, "British Petroleum," and switched to the tagline "Beyond Petroleum") will spend $8 billion on developing alternative energy over the next decade. And Wal-Mart, the world's biggest retailer, is heavily pushing its low-energy lightbulbs (as well as pledging to implement emissions-savings efforts across the company).[26]

Others want to be in on the act when it comes to setting the inevitable constraints on greenhouse emissions. For a report entitled "Getting Ahead of the Curve" the Michigan-based Pew Center on Climate Change surveyed thirty-one major American corporations.[27] Almost all of them believed that government regulation of greenhouse gases was imminent, and 84 percent believed it would happen before 2015. General Motors, Ford, and Chrysler are the latest additions to the U.S. Climate Action Partnership, which is lobbying the U.S. government to set obligatory limits on emissions.

The Pew Center report quotes Linda Fisher, vice president of DuPont, thus: "We need to understand, measure and assess market opportunities. How do you know and communicate which products will be successful in a greenhouse gas constrained world? How should we target our research? Can we find creative ways to use renewables? Can we change societal behavior through products and technologies? The company that answers these questions successfully will be the winner."

The color of money is now officially green.

11

THE ROAD FROM KYOTO

The famous Kyoto Protocol was born out of the 1992 Rio Earth Summit, the midwife being the United Nations Framework Convention on Climate Change (UNFCCC). This was the world's first climate treaty. It came into force in 1994 and has since been signed by 189 countries—including the United States and Australia—which is as near to global unanimity as any international treaty that has ever existed.[1] Its stated objective is to achieve "stabilization of greenhouse gas concentrations in the atmosphere at a level that would prevent dangerous [human-induced] interference with the climate system."

Technically, the Kyoto Protocol was an amendment to the UNFCCC treaty, assigning specific obligatory emissions cuts to each of the signatories. Any signatory who then went on to ratify the Protocol agreed to be bound by its terms. In total, the Protocol was designed to cut global emissions by 5 percent of their 1990 levels by 2012. Though this was far too low to make a significant difference to the problem, it was always intended to be a first step, after which the levels would be ratcheted upward.

Though any party to the UNFCCC could sign and ratify the treaty, only the so-called Annex 1 countries—the most industrialized nations—had any obligation to cut their emissions. A developing country that ratified would then be allowed to put forward projects for the Clean Development Mechanism, which we described in chapter 10, and through which the richer nations could satisfy some of their obligation by investing in clean technologies in the poorer ones.

Kyoto may have had its flaws, but it was still an important first step in the tortuous process of reaching international agreement to tackle climate change. As witness to that, nearly everybody signed up. Of the available 178 parties, 176 have now ratified the Protocol. Of the remaining 2, Kazakhstan intends to ratify but has not yet done so for technical reasons relating to its change in status from developing to developed country.

The only country that has not ratified, and that has declared that it will never ratify, is the United States.[2] The government has been increasingly isolated in its opinion of Kyoto in particular and agreements to cap greenhouse emissions in general. It is, however, starting to change its tune on climate change, and we will talk much more about this below and in chapter 13.

As the world's richest economy and one of the highest greenhouse emitters, the United States is particularly important; the rest of the world knows very well that the participation of the United States in a future agreement will be essential.

The first phase of the Kyoto Protocol will end in 2012, and the world urgently needs to settle on a new agreement that is much more ambitious in its targets. While the science tells us that we must reduce emissions drastically, we are presently on the wrong track. Global emissions are accelerating. In the 1990s, the decade during which Kyoto was devised and signed, greenhouse emissions from fossil fuels and industry grew at a rate of 1.1 percent per year. By the early 2000s this had risen to more than 3 percent per year and counting.[3]

Where Do I Sign?

The new international agreement will need to cover four main points:

Global Target

Everybody involved in the agreement (which is to say every nation on Earth) will have to agree on the overall global objective. Just how much are we prepared to let temperatures rise before we can stop climate change in its tracks?

Of all the decisions in the new international agreement, this will be at once the most difficult and the most important. It will make the biggest overall difference to each country's actions, regardless of how the reductions are ultimately divided among the nations. It will also make the biggest difference to our chances of averting global catastrophe.

The atmosphere already contains around 430 ppm of greenhouse gases. As we explained in chapter 6, to avoid the worst consequences of climate change we will need to end up at 450 ppm equivalent. Roughly speaking, to reach this figure industrialized countries will need to cut their emissions from 1990 levels by 80 percent by the middle of the century.[4] We will also need to set interim targets for, say, 2025 or 2030, so that we can be sure we're on the right track. For developing countries the numbers vary much more widely, depending on the way that the burden of cuts is distributed.

National Targets

Once the overall target is set, the next task will be to decide how the reductions should be divided among the nations of the world. This will be fiendishly difficult, given the need to recon-

cile the economic needs and political expectations of countries
with such wildly different backgrounds. The criteria for handing
out responsibility might include a country's greenhouse emissions per head of population; its overall emissions; its historical
responsibility for the current atmospheric load of greenhouse
gases; its wealth and economic capacity to change; and its vulnerability to near and long-term threats from climate change.

One of the hardest questions will be where to start. What
should the baseline emissions level be when deciding how much
each country needs to cut? The Kyoto Protocol used 1990 as the
baseline, which is one reason why the United States refused to
ratify it. By 1997, when President Clinton signed the Protocol,
emissions in the United States had already skyrocketed from
their 1990 level. Many American politicians felt the resulting
targets would be too draconian for the good of America, and
both Democrats and Republicans subsequently voted it down.
Meanwhile, thanks to the economic slowdown that followed
perestroika, emissions from the former Soviet bloc had plummeted relative to 1990, meaning that their emissions had also
tumbled. They could more than meet their Kyoto requirements
without doing anything at all, and any subsequent agreement
based on 1990 would leave them with billions of dollars' worth
of emissions permits to burn. Thus, though 1990 remains a point
of reference for most of the world, the former Soviet bloc would
benefit too much from it and the United States too little. (When
the G8 summit in Germany in June 2007 issued a joint statement discussing the need to halve global emissions by 2050, it
said nothing about what the starting point for this should be,
because the issue of 1990 was too controversial.)

One way around this problem that many analysts are now
adopting for a post-Kyoto agreement is to set the 2010 Kyoto
targets, which are a percentage of 1990 emissions, as a starting
point for most of the signatories, but to make actual 2010 emissions the starting point for the United States and the Russian

Federation. The disadvantage of this proposal is that the rest of the world might view it as a reward for American profligacy; the advantage is that it makes any future U.S. president much more likely to sign. This solution would cost Russia its chance to sell those billions of dollars' worth of emissions permits that arose from its economic free fall in the 1990s. But until now Russia has made no effort to sell—probably no nation other than the United States could afford to buy on that scale, and the United States is emphatically not a member of the Kyoto trading scheme. Since the United States is also most unlikely to agree to any future scheme that would keep 1990 as the baseline date, it is doubtful that these permits are of much worth to Russia in any case.

Once the target date is set, the next question is how to spread the load. First will come the important—and contentious—issue of how to incorporate aviation and shipping into the mix. Both of these sectors are projected to increase dramatically over the next few decades, albeit from a relatively low base. But it's difficult to decide how to allocate emissions from international flights and shipping: Do they belong to the initiating country, the receiving country, or the country that has most passengers or freight on board? There's also the danger that if not all countries have signed up, operators would be able to dodge regulations by fueling up in a country that's not compliant. For reasons like these, both aviation and shipping remained out of the Kyoto agreement and are not formally counted in any individual nation's greenhouse tally. Yet, as Friends of the Earth memorably put it, "a carbon management system that simply leaves these emissions out is rather like a calorie controlled diet that opts to exclude calories from chocolate."[5]

Next, and even more important, the new agreement will need to find a way to distribute the load between industrialized and developing countries. President Bush has repeatedly said that the United States will not sign any agreement that does not involve developing countries. It's true that most recent growth in emissions has come from countries outside Annex 1, but that's not the whole story. In 2004 the combined emissions of developing and least developed economies (which comprise some 80 percent of the world's population) accounted for 73 percent of emissions *growth* but still only 41 percent of total global emissions, and just 23 percent of cumulative emissions since the mid-eighteenth century.[6] Moreover, almost all developing countries have much lower emissions per person than their richer, more profligate, industrialized counterparts. (For instance, U.S. emissions per capita in 2004 were twenty-four tons of CO_2eq compared to China's meager five.)[7] And all the countries of the developing world can also argue that they want for their citizens the standard of living that Americans already have.

Also important will be to acknowledge that the world's rich countries got where they are today by exploiting cheap energy and thereby filling the atmosphere with most of its current greenhouse load. Carbon dioxide in particular lasts so long in the atmosphere that a significant proportion of the molecules generated by the steam engines and coal fires of the early Industrial Revolution are still around in the air today.

During the 1997 Kyoto negotiations Brazil made a suggestion that has since become known as the Brazilian Proposal. Its idea was that countries should now share the burden of emissions cuts according to how historically responsible they were for the problem. In other words, we should calculate what concentration of greenhouse gases each country has put into the atmosphere over time and use those figures to allocate emissions cuts. That would mean, for instance, that countries such as Germany and the United Kingdom, which have been emitting for longer than most, would bear a larger share than their current emissions implied. It would also mean that big emitters that had developed their industries more recently, such as Australia, would bear less of a share.

To date, the Brazilian Proposal has not been adopted, but it does show how imaginative thinking will be needed to develop a way of distributing the burden of emissions cuts that is clearly fair, will achieve the required target, and—most important—that will persuade everyone to sign up. The eventual agreement will have to find some way of accounting for the different state of development, and historical responsibility, of the signatories. It will have to be flexible, be cost-effective, and do the climate job. Here are some of the ideas already on the various negotiating tables, along with the buzzwords to look out for. Note that these are all ways to distribute national targets, and that they say nothing about how the individual countries will go about achieving them, or what will happen if they don't.

CONTRACTION AND CONVERGENCE (OR, "WE'LL MEET YOU IN 2050")

"Convergence" refers to a certain low target of greenhouse emissions per capita that every country agrees to converge on by, say, 2050. For instance, a global target of 450 ppm of greenhouse gases would mean convergence in 2050 at about two tons of CO_2eq per person.[8] The industrialized world would need to "contract" its emissions to meet the target, whereas developing countries would be allowed to increase their emissions and develop their economies before everyone eventually converges on the same spot. This has the benefits that every nation is involved from the beginning, that it's a transparent, straightforward concept, and that it produces a definite final concentration of greenhouse gases. However, it doesn't take into account the very different circumstances of different nations, nor their historical responsibility for our current crisis.

DIFFERENTIATED CONVERGENCE (OR, "YOU START, WE'LL FOLLOW")

This is a slightly more complex version of contraction and convergence. The richest countries would converge their emissions per head to a certain low figure within an agreed time frame. Individual developing countries would agree to converge to the same figure, but they wouldn't have to start until their emissions per head reached a certain percentage of the (steadily declining) global average. If they never pass the threshold, they never have to make cuts. This has many of the benefits of the first scheme, with the additional attraction that it puts a little more of the onus on the countries that have been historically responsible for the problem, while giving more breathing room to the developing countries trying to catch up.

MULTISTAGE TARGETS (OR, "CLIMB UP WHEN YOU'RE READY")

This approach involves a set of different stages with varying levels of commitment. Countries graduate from one to another when they pass certain thresholds (for instance, emissions per head, or GDP per head). A lot would hinge on the choice of thresholds, as well as the nature of the commitments in the later stages. One of the hazards of this approach is that it would be difficult to be sure of reaching a set global target, given the uncertainties in the different development paths. There would also have to be built-in incentives to encourage countries to act when they passed the different thresholds.

GLOBAL TRIPTYCH (OR, "DO WHAT YOU CAN")

The idea here is to allocate allowances for individual emissions-generating sectors (for instance, electricity, agriculture, or waste), then add them up to a fixed national allowance for each country. Only the overall national target is obligatory—the country can choose whatever way it likes to get there—but the process is supposed to make sure that the target is reasonably achievable. The great advantage of this approach is that it specifically takes national circumstances into account. The main disadvantage is its inherent complexity. It would require huge amounts of reliable data for each nation, it would be highly complicated to negotiate, and the burden of specificity and detail wouldn't lead to a result that was transparently fair.

SECTOR BY SECTOR (OR, "I WILL IF YOU WILL")

The idea here is to alleviate concerns about the effect of emissions reductions on global competitiveness by applying the same rules to an individual sector regardless of the country involved. So, for example, the entire world car industry could agree to a standard level of emissions per passenger mile, regardless of where the car was built or driven. Or the steel industry could agree to emit no more than a certain amount of carbon dioxide per ton of steel produced. Involving the top ten highest emitters in each sector would cover 80 to 90 percent of the emissions from developing countries.

This would certainly help deal with the worry that any company that cut emissions from its own power plants would then lose customers to ones that continued to pollute, but produced cheaper electricity. But it would not guarantee that we stay below that magic threshold of maximum greenhouse gases in the air.

INTENSITY (OR, "NEVER MIND THE TOTAL, FEEL THE EFFICIENCY")

This approach abandons the idea of overall emissions targets in favor of calculating the emissions per unit of economic growth (GDP). Its obvious benefit is that it would ensure that future economic growth is as carbon-efficient as possible, and that if future growth patterns change the target doesn't move.

The major, and perhaps overwhelming, disadvantage of this approach is that it says nothing about how much greenhouse gas ends up in the air. Even if every country in the world drastically improved its carbon efficiency, as economies continued to grow, greenhouse gases would continue to rise, and the climate would be left free to do its wicked worst. As one assessment puts it of this approach, "environmental effectiveness not guaranteed."[9] It is, however, currently much favored by President Bush.

Carrot and Stick

The third pillar of the new agreement will be an ultrareliable mechanism to encourage nations to meet their targets, and provide sanctions if they don't. Both the carrot and the stick are likely to be financial.

For encouragement, emissions trading is the best game in town. The more emissions you save, the more money you can earn by selling your excess. For sanction, the likeliest instrument is the World Trade Organization. All countries know that it's good for their economies to be part of the global trading club. In the intense horse trading that preceded the activation of the Kyoto agreement, Russia finally agreed to sign up as a straight swap for European support of its WTO membership application. Since the WTO sets global trading rules, it will also need to ensure that every country accepts that economic benefits go hand in hand with meeting climate obligations, and that the world won't tolerate, and can't afford, climate freeloaders. Public opinion could also play an important role in the individual countries. The *Stern Review* puts this nicely: "As the science of climate change is widely accepted, public attitudes will make it increasingly difficult for political leaders around the world to downplay the importance of serious action."[10] We'll say more about this in chapter 14.

Flow of Resources

Finally, the agreement will need a mechanism to transfer both technology and funds from the richest countries to the developing world.

TECHNOLOGY TRANSFER

Developing countries will need to be able to continue their economic growth without pouring out carbon dioxide, by leap-frogging over the polluting technologies that got the first world where it is today. It's in nobody's interest if countries don't meet their targets because they can't afford to do so. And no developing countries will come on board and agree to curb their emissions unless some kind of technology transfer is in place. It is likely to be based on some form of Clean Development Mechanism, driven by a carbon-emissions-trading market. Finally, this part of the agreement will need to include some way to pay for preserving forests through cutting deforestation rates, which is something that Kyoto didn't do.

ADAPTATION FUNDING

As we've repeatedly stated, the science of climate change makes it very clear that the countries least responsible for causing the problem will be the ones to suffer most. They are also the countries least able to pay for the sorts of adaptations that are already necessary. At a pinch, and with the right political will, the United States could protect its coastline from another Katrina-type flooding disaster, and the United Kingdom is already protecting its own coast. But the same doesn't apply to Bangladesh, which, as well as having the potential to be harder hit, has a much weaker economy. The international agreement will have to deal with this issue by ensuring the flow of adaptation funding from rich to poorer nations.

Both technology transfer and adaptation funding will probably go hand in hand. For instance, it could be part of the agreement that if a rich nation exceeds its agreed-upon emissions it can either buy unused permits from elsewhere or provide technological or adaptation costs.

Last-Chance Saloon

It's unlikely that all the details of such an agreement can be thrashed out among the 189 countries that have signed the Kyoto Protocol. For such delicate negotiations, this group is just too unwieldy. However, there is more hope to be found in a new grouping of the topmost polluting nations. This group, now enshrined as the Gleneagles Dialogue, comprises the G8 richest nations, the European Union, and—in particular—the top five most rapidly developing economies: the so-called G8 + 5. If these very disparate countries, the world's major emitters, can be brought to agreement, the rest of the world might well follow. The next two chapters will look at these key countries in more detail.

Above all, the new agreement needs to come soon. It's not just that the Kyoto Protocol runs out in 2012. More important, we are running out of time to come to grips with climate change. If we had acted three decades ago we might have been able to prevent the dangerous effects that we are witnessing today and will continue to witness, come what may, over the next few decades. Now it's too late for that.

But we still have time to stop the worst of the possible future consequences. As we've already emphasized, that will probably require emissions levels to peak within the next fifteen years, before starting to fall—which in turn means we need to act fast. Given how long it takes to set the wheels of government turning, even with the best will in the world we must have an international agreement in place as early as possible in 2009.

The key international meeting to look out for is the UN Climate Summit in Copenhagen in December 2009. All the signatories of Kyoto will be present, and this is when the essential elements of the new treaty will finally have to be decided.

When it comes to climate change there's no more time for fooling around. This is the last-chance saloon.

12

RAPIDLY DEVELOPING NATIONS
(or, come on in, the agreement's lovely)

Tackling climate change will require the cooperation of the entire world, but some countries will play a greater role than others. In particular, the most important changes will need to come from two sets of players: the industrialized world, which has the richest economies and bears the greatest historical responsibility for the emissions to date; and the handful of most rapidly developing countries that are likely to contribute the most to future rises in emissions as they play catch-up.

Perhaps it's human nature to forget what's gone before and focus on what's to come, or perhaps it's a convenient escape for those of us who live in industrialized countries. Either way, the latest fashion in the industrialized world is to declare that there is no point in reducing emissions, since any reductions will be swamped by the vast increases to come in rapidly developing countries such as China and India. Because of this, we have decided to start with the five countries that are currently developing most rapidly.

That doesn't mean we believe these countries are "the problem." It's true that they are responsible for most of the recent spurt in greenhouse gas emissions, and that they make up a substantial proportion of the global total. However, most of them are also acutely aware of the dangers likely to come from climate change. In fact, as you'll see, these countries are already going to great lengths to deal with their emissions, often under very difficult circumstances.

Moreover, and most important, these countries tend to produce only very small amounts of greenhouse gas per individual citizen and have been responsible for very little of the current concentration of greenhouse gases in the atmosphere. They also have this in common: Their development needs tend to swamp all other political and economic considerations.

Nonetheless, the five most rapidly developing nations that we describe here will be vital to any future agreement. If any one of them stays outside the agreement, that country could then become a hothouse manufacturer, producing the goods—and the greenhouse emissions—that the other nations weren't allowed to produce, so that there would be no net savings in global emissions.

The numbers given for emissions for each country come from a report commissioned by the British government from a German consulting firm called Ecofys.[1] Different sources can give confusingly different figures, which are further confounded when some figures are only for carbon, some only for carbon dioxide, and some for all greenhouse gases, calculated either as carbon equivalents or as carbon dioxide equivalents. The figures given here for emissions are for all greenhouse gases, calculated as a carbon dioxide equivalent. Emissions per capita and total emissions are for the year 2004. Historical emissions are average annual values from 1900 to 2004, per head of the present population.

China

Per capita: 5.0 tons/person
Historical: 1.2 tons/person
Total: 6,467 megatons
Change in emissions since 1990: +72.7 percent
Ratified Kyoto? Yes

China has become the latest, greatest bogeyman for climate change skeptics, so it's appropriate to start here. The pace of development in the country is certainly extraordinary. A year or so ago, China was building a new coal-fired power station every week; now it's closer to two a week and counting. China has no oil reserves and very little gas. But it does have coal, and plenty of it. And that coal is firing the surge in China's economy.

It has become fashionable to quote the figures for China's new power stations, and to use them to argue that there's no point in Western countries doing anything to halt carbon emissions. These new power stations are especially bad news from a climate perspective because coal is the dirtiest of all the fossil fuels, producing not just smoke and smog in the cities, but also much more carbon dioxide for every unit of energy than either oil or gas. Moreover, although the figures given here from 2004 show China's total output lagging slightly behind that of the United States, by one measure China has now outstripped the United States to take the uncoveted position of the world's biggest greenhouse polluter.[2]

Is it fair, then, to say that China is the biggest greenhouse problem? Not exactly. For one thing, many of the countries in the Western world have dodged their own carbon dioxide emissions by exporting their manufacturing to . . . China. Next time you buy something with MADE IN CHINA stamped on it, ask yourself who was responsible for the emissions that created it.

Even more important, China is one of the smallest emitters when you calculate emissions per capita, especially compared to industrialized nations, and its relative historical contribution to the problem is truly tiny. China can say with justice that, unlike the industrialized West, it has done almost nothing to create the climate problem, and that its citizens play on average a very meager part in perpetuating it.

The Chinese government's biggest priority is the massive disparity in wealth between the rich citizens of Beijing and Shanghai, and the seven hundred million or so people living on less than $2 per day. Bridging this gap to bring at least a decent standard of living to China's vast population, which is what's behind the plethora of power stations, is surely a reasonable goal in anyone's book.

However, unless China finds a way to develop without massively increasing its greenhouse emissions, the efforts of the rest of the world will count for very little. For the new international agreement on emissions to stand any chance of success, China has to be on board. (The same, by the way, also applies to the United States—see below.)

The good news is that the Chinese government is at least as aware as any other country of the dangers of global warming. Unlike most other governments, fully two-thirds of the members of its politburo are highly qualified scientists and engineers. These are people who fully understand the climate problem.

They are aware, for instance, that China itself would be hard hit by unreconstructed climate change. There is already a shortage of irrigated land in the interior, and this will only be made worse as the Tibetan glaciers shrink and the rivers they feed dry up. And Shanghai, the country's economic powerhouse, is one of the world's major cities most vulnerable to flooding both from inland rivers and the rising sea. The Chinese government has already signed an agreement with the United Kingdom to investigate ways to improve Shanghai's flood defenses, but unless climate change is held in check it will be a real struggle to keep the city above water by the end of the century.

When it comes to a new agreement, China will need to be allowed some expansion of emissions, perhaps through a contraction-and-convergence approach, along with financial investments from industrial nations through some kind of Clean

Development Mechanism. For instance, since the coal-fired power stations are inevitable, it's vital to find a way to capture the carbon dioxide emissions and store them away from the air. The European Union has recently agreed to joint funding of a Chinese pilot program for carbon capture and storage, and the United Kingdom has also agreed to a project to map potential Chinese sites for burying the carbon. This is especially urgent because retrofitting those two new power stations a week will be much more expensive than building in the design from the start.

But the most important factor in bringing the Chinese into any agreement will be participation by the United States. As long as U.S leaders refuse to agree to targets, the Chinese will be able to cry "hypocrite." China will come on board when the United States does, and using the most reasonable measures of carbon emissions responsibility—historical and per-capita emissions—China definitely holds the moral high ground.

Brazil

Per capita: 5.3 tons/person (plus approx. 7 tons/person from deforestation)
Historical: 1.6 tons/person
Total: 983 megatons (plus approx. 1,100 megatons from deforestation)
Change in emissions since 1990: +40.7 percent
Ratified Kyoto? Yes

Brazil needs no persuasion of the gravity of the climate change problem. Back in 1992 it offered Rio as the venue for the first international Earth Summit, which launched the process that led to the Kyoto Protocol. The government is eager to involve developing nations in the solution. It is also highly aware of the potential threat that climate change poses to Brazil, especially reduced food yield, forced changes in land use, and loss of forest. Brazil's agriculture depends critically on the weather and is highly sensitive to changes in temperature and rainfall. Brazil also has excellent climate scientists, to whom the present government clearly listens.

Despite a relatively high state of development, especially in the south of the country, Brazil still has a low figure for its greenhouse emissions per capita. One reason is that it uses hydropower for much of its electricity, giving it one of the world's lowest power emissions rates. Also, after the 1970s oil crisis Brazil began using sugarcane to make alcohol for vehicle transportation. It is now the acknowledged world leader in use of biofuels, with extremely low transportation emissions: 0.74 tons of carbon dioxide equivalent per head, compared to 2.24 for the United Kingdom and 6.36 for the United States. Overall renewable resources account for 40 percent of its energy supply.

However, deforestation in the Amazonian north and the Atlantic northeast of the country still produces huge amounts of carbon dioxide. Most sets of greenhouse figures count only emissions from the use of fossil fuels, and in Brazil's case adding in the figure from deforestation makes a huge difference. Brazil's government is aware of this and in the past few years has taken significant steps to expand the range of protected areas and prevent illegal logging. More than 40 percent of the Brazilian Amazon is now under some form of protection.[3]

The problem is that there is at present virtually no economic incentive to prevent deforestation. Under the Kyoto Protocol, industrialized countries can pay Brazil to *reforest* land to balance their own excess emissions, but stopping deforestation was considered too problematic and stayed out of the Clean Development Mechanism. Any future proposal would need to make forest protection a part of the mix.

Considering the rest of Brazil's emissions, contraction and

convergence would require slowing the rate of emissions growth by 2020 and reducing overall emissions by 2050. The multistage approach would allow a little more short-term breathing space, and the sectoral approach would be especially good for Brazil since its electricity emissions are already so low.

South Africa

Per capita: 11.1 tons/person
Historical: 3.6 tons/person
Total: 505 megatons
Change in emissions since 1990: +29.7 percent
Ratified Kyoto? Yes

For a developing country, South Africa has an unusually high level of per-capita greenhouse emissions. Like China, its economy is heavily dependent on coal, but the country is considerably more developed. The mining industry—for gold, diamonds, platinum, and uranium—is a major part of the economy and is very energy intensive. As the gold has begun to run out, the country has been taking on manufacturing commissions from developed countries. For instance, all BMW 3-series cars are now assembled at Pretoria's Rosslyn plant.

South Africa thus has a large amount of energy-intensive industry, and that looks set to continue. It will also continue to be a mining economy for the foreseeable future. As we mentioned in chapter 9, the country is investing in the development of Pebble Bed Nuclear Reactors, but this is likely to be for export rather than signaling a major nuclear strategy.

The country also has a well-developed sugarcane industry and ample technological skills for converting a proportion of this to biofuels, although it would be hard to expand the current areas and grow more. It is also still wrestling with the aftermath of apartheid, as well as the ever-present trauma of AIDS.

The current government is nonetheless progressive on the issue of climate change and eager to talk, at least at the level of energy and environment ministers (though less so at the level of heads of state). Overall, the government is determined to be a responsible international player. If the terms for developing countries were right and allowed the country to deal with its myriad social and developmental issues, South Africa would participate.

Mexico

Per capita: 5.0 tons/person
Historical: 1.3 tons/person
Total: 520 megatons
Change in emissions since 1990: +38.6 percent
Ratified Kyoto? Yes

As with the other rapidly developing countries, Mexico's first priority is growth. However, the government is also highly aware of the threat of climate change, through desertification, loss of food productivity, and coastal flooding from rising sea levels and intensifying tropical storms.

Mexico is already beginning to make some efforts to slow the increase in its greenhouse emissions. The government is also eager to be engaged with the G8 + 5 Gleneagles Dialogue. In 2006 Mexico hosted the first follow-up meeting of this group after it was initiated the previous year at Gleneagles. If it could be persuaded that its immediate economic growth were not at risk from the new treaty, and if there were a sense of common purpose within the G8 + 5 group, Mexico would be a very good player.

India

Per capita: 1.6 tons/person
Historical: 0.6 tons/person
Total: 1,744 megatons
Change in emissions since 1990: +57.5 percent
Ratified Kyoto? Yes

India is probably the most distracted of the rapidly developing countries from the climate point of view. It has by far the lowest per-capita emissions level of the group, and also by far the most pressing developmental problems.

However, the subcontinent as a whole stands to suffer dramatically from global warming. Food production is forecast to drop, and the Ganges–Brahmaputra megadelta in Bangladesh is frighteningly vulnerable to flooding from a combination of sea-level rise and more intense tropical storms. And although the models are as yet unsure about how the Asian monsoon will change with increasing temperatures, nobody expects it to stay as it is.

Indian scientists are certainly well aware of this—the chairman of the IPCC, Rajendra "Pachy" Pachauri, is based in Delhi, and the Indian National Academy of Sciences has signed three successive global statements calling for action on climate change. India has also already installed a huge amount of wind power, and an Indian company, Suzlon, is one of the world's leading manufacturers of wind turbines.

But for the moment the urgency of the problem has not fully penetrated the Indian government. Still, given its relatively low state of development but high rate of growth, India is one of the most fruitful sources of opportunities for projects involving the Clean Development Mechanism. Moreover, an encouraging sign came recently from Kapil Siba, the minister for science and engineering. Siba recently requested a meeting with one of us (David King) specifically to propose that India set up an Energy Technologies Institute to develop new, low-carbon energy sources in parallel with the British one. This was the first official thawing of the Indian government's position on climate change.

If the new agreement allows India to continue its urgent development needs and potentially profit from international investment via some means similar to the Clean Development Mechanism (through which it has already received major international investment), it is very likely to get on board. Since it is starting from such a low emissions rate per capita, contraction and convergence would allow India to keep its emissions growing with little interference over the coming decade, while bringing financial benefits from the sales of emissions allowances.

13

INDUSTRIALIZED NATIONS
(or, whose fault is it anyway?)

The world's major industrialized countries will clearly need to take the lead in tackling climate change. Together they have been responsible for almost all of the current climate problem; they gained their wealth and advanced state of development largely by exploiting cheap fossil fuels at an early stage. These are the countries that both bear the brunt of the collective responsibility for climate change and have the economic resources to tackle it. All have embraced this responsibility in principle. Most, but not all, have also begun to act.

(As in chapter 12, the figures given here for emissions are for all greenhouse gases, calculated as a carbon dioxide equivalent. Emissions per capita and total emissions are for the year 2004. Historical emissions are average annual values from 1900 to 2004, per head of the present population.)

United States

Per capita: 24.0 tons/person
Historical: 12.7 tons/person
Total: 7,065 megatons
Change in emissions since 1990: +15.7 percent
Ratified Kyoto? No
Kyoto target for 2012: −7 percent (not adopted)

The United States stands head and shoulders above the rest of the world in both its wealth and its greenhouse gas emissions. Though according to one report it has recently ceded its title as top overall emitter to China, the United States still emits vastly more per person than almost every other country. It also bears significant historical responsibility for the greenhouse gases already in the air.

Until relatively recently, the United States was also a world leader on tackling climate change. American scientists have long pioneered the study of climate change and were among the first to alert the rest of the world to the problem. The United States chaired the working group that gave rise to the first global climate treaty, the UNFCCC. And throughout the 1990s U.S. administrations were very open to finding ways to limit emissions, under the strong advocacy of Vice President Al Gore.[1]

But in recent years the United States has also, famously, pushed the issue of climate change to the sidelines (despite the best efforts of former British prime minister Tony Blair to persuade President Bush otherwise) and refused to ratify the Kyoto Protocol.

As we mentioned in chapter 11, the seeds of the American problem with Kyoto lay in the date set by the treaty as a baseline for emissions. By 1998, when President Clinton signed the treaty, American emissions had skyrocketed compared to their 1990 level, so that any agreement using 1990 as the baseline had become much harder to achieve. This led to jitters among politicians who were afraid of damaging an increasingly fragile economy. The Senate—Republicans and Democrats alike— had already voted unanimously for the Byrd–Hagel Resolution, which declared that no protocol should be signed that "would result in serious harm to the economy of the United States." The resolution also criticized the lack of commitments in Kyoto by developing countries. Thus the Clinton/Gore administration did not submit Kyoto to the Senate for ratification.

But the real blow came with the advent to power of President Bush and his fiercely partisan advisers. The issue of climate change suffered politically from its close association with the Democrats, especially Vice President Gore. Moves to change the composition of the scientists on the Intergovernmental Panel on Climate Change (IPCC), for instance, began almost immediately, with any that were considered too closely in tune with the previous administration viewed with suspicion. Shortly after the Republicans came to power, a notorious leaked memo sent from oil company ExxonMobil to the incoming administration had as its chief recommendation: "Restructure the US attendance at upcoming IPCC meetings to assure *none of the Clinton/Gore proponents are involved in any decisional activities* [their emphasis]."[2] The same memo called specifically for World Bank chief scientist Bob Watson (whose scientific credentials were impeccable, but who had previously worked with the Clinton/Gore administration) to be removed from the chairmanship of the panel. This subsequently occurred.

It probably didn't help that many of President Bush's cabinet were closely associated with the fossil-fuel industry. Before their appointments, Dick Cheney had been CEO of Halliburton; Bush's first commerce secretary, Donald Evans, had been chief executive of a leading oil and gas exploration company; and Condoleezza Rice was a director of Chevron.[3]

Efforts to halt climate change also suffered from their association with the environmental movement. To many conservative Republicans, environmentalists were scarcely better than communists for their apparent love of regulating industry and restricting economic growth. One of President Bush's first acts in power was to declare that the United States would not ratify Kyoto, and Rice famously added, "Kyoto is dead." (Reports of its death, as Mark Twain remarked on reading his own premature obituary, turned out to be greatly exaggerated. Though Bush succeeded in persuading Australia to hold off ratifying, Russia's eventual ratification took the number of countries over the required threshold, and Kyoto has been fully operating—alive and well—since February 2005.)[4]

Since then, the story of climate change in the United States has been a woeful one, with legions of top-notch American scientists calling loudly for action and the administration sometimes ignoring and sometimes—infamously—attempting to suppress these calls. (The attempts at suppression led to widespread condemnation by U.S. scientists[5] and scathing hearings in March 2007 before the Congressional Committee on Oversight and Government Reform.)[6] Jack Marburger, the president's chief science adviser, has had a tough time of it. In spite of the overwhelming evidence that dangerous climate change is already here, he waited until late 2007 before acknowledging in public that climate change is a real and human-made problem.

The oil industry in the United States has long marched in step with the administration's position—especially Exxon-Mobil, which funded a host of think tanks dedicated to rubbishing the science of global warming. One such think tank offered $10,000 each to scientists who would write an essay criticizing the findings of the IPCC report. Another—the Competitive Enterprise Institute—put out an infamous nationwide series of TV commercials suggesting that carbon dioxide was an unalloyed benefit for humankind and had been unfairly besmirched. The scientist whose findings were quoted in the commercials later angrily accused the makers of selective reporting and described the result as "a deliberate effort to confuse and mislead the public about the global warming debate."[7] (As we explained in chapter 1, carbon dioxide is indeed necessary for life, but the science of global warming shows unequivocally that you can have too much of a good thing.)

To quell the climate protagonists and to satisfy his requirement that developing countries be involved in any agreement, President Bush set up what was widely perceived as a rival to the Kyoto Protocol: the Asia-Pacific Partnership on Clean Development and Climate, involving the United States, Australia, South Korea, China, India, Japan, and Canada, which joined in October 2007. Like Kyoto, the partnership is at least outwardly dedicated to solving the climate problem. Unlike Kyoto, however, it sets no mandatory targets or specific mechanisms to achieve the voluntary targets put forward by its members. (Since the targets are not mandatory, there is also no sanction for failure to achieve them.)

The partnership centers around the transfer of low-carbon technology from developed to developing nations, with a heavy focus on carbon capture and storage—the as-yet untested method by which the carbon dioxide from coal-fired power stations can be captured and stored underground instead of being released into the atmosphere. This is likely to be a vital tool in reducing emissions from countries like China, India, and South Africa, which depend heavily on coal for their current energy sources. But on its own it's not enough.

In April 2006 the Climate Institute of Australia issued a progress report on the Asia-Pacific Partnership, which said that even in the most optimistic scenarios the agreement would lead to a global *doubling* of greenhouse gas emissions from present levels by 2050, rather than the drastic reductions that are needed to quell climate change. It also said that the amount of money invested in developing the new technologies so far is much too low to have any serious impact—just 1.1 percent of the value invested in global carbon markets as a result of the Kyoto Protocol, and a vanishingly small fraction of the amount of money the International Energy Authority says will be invested in energy over the next twenty-five years.[8] Republican senator John

McCain summed up the thoughts of many when he said that the partnership was "nothing more than a nice little public-relations ploy."[9]

However, the tide seems to be turning—in the rest of the United States, if not in Washington. Whether or not Hurricane Katrina can be attributed directly to global warming, it certainly served as a wake-up call for how dramatically nature can bite back even within a "safe" urban environment, and how foolish it can be to ignore the warning signs.

Meanwhile, Al Gore's passionate association with the cause of climate change switched from being Republican anathema to major asset after his phenomenal success with the book and subsequent movie *An Inconvenient Truth*. (Gore's winning of the 2007 Nobel Peace Prize with the IPCC for his work on climate change also helped to boost the profile of his campaign.) Another significant driving force is the state of California, which has a proud record of dealing efficiently with environmental problems—especially the once smog-choked cities of Los Angeles and San Francisco—by briskly applying laws and encouraging the development of new technologies. There was nothing sentimentally green about California's action to bring catalytic converters to the exhausts of its cars, nor did it harm the economy. Every new squeeze on the regulatory system for car exhausts met opposition from the motor industry—but they never stopped selling cars. The car-exhaust catalysts championed by California are now standard issue throughout the industrialized world, and are the main reason why so many cities are so much more pleasant places to live.

California, under the unlikely figure of its "Governator," Arnold Schwarzenegger, has now begun to tackle the climate problem with the same brisk efficiency. At a meeting of West Coast governors in 2005, Schwarzenegger invited one of us (David King) to give a speech outlining the United Kingdom's position on climate change. When King said that the United

Kingdom had committed unilaterally to reducing its emissions by 60 percent by 2050, Schwarzenegger's response was "we'll reduce by 80 percent." That's the sort of bidding war we should all be waging.

The state has begun putting its money where its mouth is. In 2006 California passed the Global Warming Solutions Act,[10] which commits the state to return to 1990 emissions levels by 2020 (which means an effective cut of 25 percent). The Act also instigated a statewide emissions-trading scheme, and Schwarzenegger himself has taken to driving a hybrid Hummer.

Nine northeastern and mid-Atlantic states have set up their own trading scheme, known as the Regional Greenhouse Gas Initiative (or, more affectionately, ReGGIe).[11] And hundreds of mayors hailing from cities in every state have signed the U.S. Conference of Mayors' Climate Protection Agreement, which pledges their home cities to reduce carbon dioxide emissions to 7 percent below 1990 levels by 2012.[12]

The climate for change has looked more promising still since Democrats regained control of both the House of Representatives and the Senate in fall 2006. The outgoing Republican chair of the Senate's influential Environment and Public Works Committee was Senator James Inhofe, who once declared that global warming was the "greatest hoax ever perpetrated on the American people." The incoming Democratic chair of the same committee, Senator Barbara Boxer, has openly welcomed the arrival of a slew of bills aimed at encouraging nationwide emissions reductions.

However, that's not to say that climate change has remained a partisan issue. Most of the new bills have both Democrats and Republicans as sponsors. And the Christian right, once fierce supporters of President Bush's position, have begun to back action on climate change on the basis that we need to act as effective stewards of God's Creation.

The private sector, too, has begun to get in on the climate

act. Companies of the stature of Wal-Mart and General Electric have converted to the climate cause, and even ExxonMobil has softened its stance. While it is still vehemently opposed to the Kyoto Protocol, it has canceled the funding of the more egregious lobby groups and begun to discuss what American carbon regulation might ultimately look like.[13]

There is new talk of climate change as a market failure born of unfair fossil-fuel subsidies, and cap-and-trading schemes as an appropriate capitalist adjustment rather than a tree-hugging regulatory tax. The myriad separate emissions-trading schemes that are springing up around the country are encouraging companies to push for a national scheme so that they can plan the future based on a single, consistent set of rules and market conditions instead of having to work with different schemes in different states. Environmental activist Robert F. Kennedy Jr. was moved to write in *Rolling Stone* that, because it encourages efficiency, "a truly free market is the planet's best friend."[14] The *New York Times* declared in December 2006 that the color of the year had been green. And the following January, in his State of the Union Address, President Bush famously declared that a way had to be found to break America's addiction to oil.

There is still, however, a long way to go. Though President Bush has begun to change his language about the dangers of climate change, his favored strategies to counter it are at best doubtful in their efficacy. They include a massive increase in the use of biofuels—which, as we explained in chapter 8, comes with its own dangers—coupled with an effort to reduce the intensity of carbon emissions per unit GDP by 18 percent by 2012. This sounds impressive until you discover that the overall emissions would still rise by 13 percent, which is only a whisker away from the do-nothing "business as usual" scenario.[15]

Moreover, though the president's negotiators agreed to the joint G8 statement issued in Germany in 2007, which accepted the reality of climate change and agreed to work toward reducing global emissions, a fascinating document leaked by Greenpeace showed the extent to which those same negotiators had insisted on watering down the statement's language. For instance, they refused to allow any mention of binding commitments, saying that this approach was "fundamentally incompatible with the President's approach to climate change."[16] One lobbyist accused the Bush administration of "trying to lay landmines under a post-Kyoto agreement after they leave office."[17]

Just before the meeting in Germany, President Bush proposed to set up talks, chaired by the United States, between the world's most polluting nations, to run "in parallel" with the United Nations efforts. However, this was shouted down by other nations, especially Brazil, who saw it as an attempt to undermine the post-Kyoto process. President Bush then reluctantly assented to the statement put out by German chancellor Angela Merkel, who was chairing the G8 + 5 meeting, that "[w]e have agreed that the UN climate process is the appropriate forum for negotiating future global action on climate change."

Given this record of recalcitrance followed by reluctance, it might be too much to hope that the current administration will fully embrace effective action on climate change. But the groundswell of grassroots activity, coupled with the shifts in attitude in American business and on Capitol Hill, might well be enough to persuade the next presidential incumbent—whether Republican or Democrat—that action is very much in the national interest. They won't have far to go. Of all the prospective candidates, John McCain, Chuck Hagel, Barack Obama, John Edwards, Mitt Romney, and Hillary Clinton already take a much more progressive stance on climate change than the outgoing president.

Nonetheless, the United States is one of the world's largest emitters of greenhouse gases per person, is the second-largest emitter overall, bears an enormous responsibility for historical emissions, and also has the world's most advanced economy. Though the rest of the world is not sitting on its hands, the new global treaty will have little chance of success until and unless the United States takes a position of responsibility and global leadership on the issue of climate change.

Russian Federation

Per capita: 13.5 tons/person
Historical: 7.1 tons/person
Total: 1,938 megatons
Change in emissions since 1990: −35.1 percent
Ratified Kyoto? Yes
Kyoto target for 2012: 0 percent

In the run-up to the Kyoto agreement, the Russian government repeatedly declared that climate change is very far from its priorities. President Vladimir Putin nailed his colors to the mast in 2003 when he said, "Russia is a northern country and if temperatures get warmer by two or three degrees, it's not such a bad thing. We could spend less on warm coats, and . . . grain harvests would increase."[18] Whether the grain harvest in Russia would indeed go up with a warming of three degrees remains debatable. But every model agrees that above three degrees the harvest would plummet. And climate change would bring other troubles to Russia that might well outweigh the benefits of fewer fur coats.

By 2070, for instance, the IPCC report predicts significantly increased flooding in northern Russia and more frequent droughts in the south of the country. It also predicts that melting permafrost will not just lead to buckled and broken gas pipelines as the ground falls beneath their feet, but will also expose vast amounts of drained wetlands in Russia to catastrophic fires similar to the ones that swept Borneo and smothered the country in a fug of orange smoke.[19] The problem with climate change is that it's all or nothing—you can't cherry-pick the blessings and leave aside the curses.

Putin made this speech while he was considering whether to ratify Kyoto. The decision was significant because the Protocol was designed not to come into effect until countries responsible for at least 55 percent of the 1990 carbon dioxide emissions had agreed to go ahead. By 2003, of all the big industrialized polluters only Russia, Australia, and the United States were yet to ratify. Any one of the three would take the count over the 55 percent threshold, but if none of them ratified, the Protocol would be doomed.

The United States had already declared its intention not to ratify, and after intense pressure, Australia also refused. Meanwhile, the signs from Russia were not good. A bilateral meeting on climate change between British and Russian scientists was effectively hijacked when Putin's then chief of staff, Andrei Illarionov, changed the agenda at the last minute to bring in a ragbag of skeptics and attempt to prevent the British scientists from speaking.

One of us (David King) was leading the UK delegation. He had already explained that he would have to leave during the conference for further government meetings at the Kremlin. But when he moved to depart, Illarionov seized the microphone and began screaming that he was a coward and "commanding" him to stay put to answer his questions. The British delegation left with this extraordinary outburst still ringing in their ears. The next morning's *Moscow Times* bore the unpromising headline THE UK DECLARES WAR ON RUSSIA, which turned out to be a direct quote from Illarionov.

But where scientific arguments cannot sway a government, economic ones sometimes prove more powerful. Russia was eager to join the global trading club, the World Trade Organization. Following David King's report back from Russia, Tony Blair went to the EU leaders to propose a horse trade. If Russia ratified Kyoto, the European Union would support its entry into the WTO. This proved a powerful trump card. On November

4, 2004, and in the teeth of Illarionov's continued opposition, President Putin ratified the Kyoto Protocol.

To Russia, backing Kyoto made little practical difference. As previously mentioned, for purely economic reasons its emissions had plummeted after 1990, meaning that it could meet its Kyoto obligations with ease. In fact, its emissions had dropped so low that it was holding billions of dollars' worth of permits that, in principle at least, it could sell on the Kyoto trading market. However, it didn't do so—probably because only the United States could have afforded to buy them, and the United States remained aloof from Kyoto.

Russia's support for a post-Kyoto treaty remains uncertain. On the one hand, its new isolationism coupled with its considerable natural gas exports suggest that it will want to hold on to a carbon-based economy. But on the other, its potential for selling at least some of its permits might well be tempting enough to keep Russia on board if the United States is part of the mix. However, as we explained in chapter 11, any future agreement would almost certainly have to reset Russia's starting level to a more recent emissions figure than 1990.

Japan

Per capita: 10.6 tons/person
Historical: 4.0 tons/person
Total: 1,355 megatons
Change in emissions since 1990: +8 percent
Ratified Kyoto? Yes
Kyoto target for 2012: −6 percent

Considering its advanced state of industrial development, Japan has relatively low emissions per head, mainly because much of its electricity comes from nuclear power, and its industry is one of the most energy efficient in the world. It has also long been committed to tackling climate change, being the host country of the meeting that created the Kyoto Protocol.

Partly because of this, and also because Japan has no oil reserves of its own, saving energy has been elevated to the status of a national imperative. In 2005 the Japanese government's "cool biz" campaign encouraged civil servants to set their office air conditioners to no lower than 28°C (82.4°F). Thousands of workers agreed to remove their suit jackets and ties and switch to short-sleeved shirts for the summer—and this in a conservative and conformist business-oriented culture. In metropolitan Tokyo alone, the campaign saved enough energy to power a city of a quarter of a million people for a month.[20] The campaign has now become an annual event. At the beginning of each summer since then, the prime minister has removed his jacket and tie, and his cabinet and an extraordinary proportion of the country have followed suit.

The Japanese government remains nervous about the effects that reducing carbon emissions further may have on the Japanese economy, and the country is focusing on technological innovation. It was one of the founder members of the new international fusion project that we described in chapter 9. It is also the world's biggest investor in solar photovoltaics and in techniques to make hydrogen gas from water.

Any future approach based on per-capita emissions would be favorable to Japan, since this would lend weight to the high level of efficiency it has already achieved.

Canada

Per capita: 23.7 tons/person
Historical: 9.8 tons/person
Total: 758 megatons
Change in emissions since 1990: +27 percent

Ratified Kyoto? Yes
Kyoto target for 2012: −6 percent

Canada ratified Kyoto in 2002, but it is presently very far away from achieving its target of a 6 percent cut in emissions. In fact, between 1990 and 2004 its emissions increased by 27 percent, which is proportionally even more than for the United States.

There is a great deal of popular support in the country as a whole for action to deal with climate change, but successive governments have stalled on how to put this into practice. In 2005 the then government published a proposal for reaching the Kyoto target using a combination of mandatory emissions cuts for large factories and power plants and largely voluntary schemes for other sectors. Environmental groups criticized the plan for relying too heavily on voluntary measures and incentives rather than obligatory cuts, but in any case the government did not implement the plan before it was voted out of office in 2006.

The current Conservative minority government, under Prime Minister Stephen Harper, then declared that Canada's Kyoto target was now impossible and cut the funding for many climate-focused schemes. The government did produce an alternative plan to tackle emissions, but this was muddied since it bracketed greenhouse gases with smog and other more immediate and apparent forms of pollution. Also, the obligatory targets set by the plan for industry to reduce its greenhouse emissions were based on their intensity—that is, emissions per unit of production—rather than their overall quantity. (We discussed in the section on the United States, above, the ways in which this approach is flawed.)

These events sparked a flurry of political activity, including the passage through Parliament of a private member's bill, "an Act to ensure that Canada meets its global climate change obligations under the Kyoto Protocol," which was designed to try to force the government to comply with Kyoto. All members of the

three opposition parties voted for this bill, and it became law on
June 22, 2007.

Among other things, the Act obliges the government to
specify the greenhouse gas reductions expected for each year
until 2012, showing how these will meet Canada's Kyoto obligations. However, in spite of several subsequent legal challenges
the sitting government has repeatedly said that it has no intention of attempting this, and at time of writing (January 2008)
the situation was still unresolved.

Australia

Per capita: 26.2 tons/person
Historical: 10.3 tons/person
Total: 529 megatons
Change in emissions since 1990: +25.9 percent
Ratified Kyoto? Yes
Kyoto target for 2012: +8 percent

Australia participated in the Gleneagles Dialogue in Mexico, and
recent droughts have prompted considerable support in the country as a whole for action on climate change. However, when he was
prime minister, John Howard steadily refused to ratify the Kyoto
Protocol. His critics say that this was mainly to curry favor with
the United States, a view that is supported by the fact that, out of
178 parties to the agreement, only the United States and Australia
have refused to ratify. Moreover, had Russia not been persuaded
to change its position and ratify in 2004, Australia's refusal would
have been enough to keep the treaty from coming into force,
which was the declared wish of President Bush's administration.

Though Australia's Kyoto target was set at a generous +8
percent above 1990 levels, its current emissions have risen still
further, by nearly 26 percent. It has one of the highest per-capita
levels of emissions in the world, higher even than the United

States. It also stands to lose heavily from future climate changes, including increased droughts and heat waves, further tropical storm damage, and the bleaching of corals on the Great Barrier Reef. (Indeed, the state of Victoria is already reeling after seven successive years of drought, and counting.)

During his time in office, Prime Minister Howard put all his faith in the Asia-Pacific Partnership on Clean Development and Climate. This sets no binding emissions targets, focuses on attempting to develop new technologies, and, as we explained in the section on the United States, by itself would lead to a significant global rise in greenhouse emissions rather than the reduction that we need.

However, the federal government was put under increasing pressure to implement stringent measures to tackle climate change. This came in part from individual state leaders. New South Wales has had an emissions-trading scheme in place for electricity since 2003, and in 2004 all the Australian states united to form a task force, called the National Emissions Trading Task Force, to push for a nationwide scheme.[21] There has also been a groundswell of popular support. In one poll 80 percent of respondents said they wanted the government to do more to tackle climate change.

In November 2006, in a remarkable turnaround, former prime minister Howard made a speech saying, "Australia will be part of future discussions which are designed to get total international agreement involving all of the major polluters, involving all the nations of the world," and mentioned, at least, "talking about an effective worldwide emissions-trading system."[22]

On June 4, 2007, following a report from the Prime Ministerial Task Group on Emissions Trading,[23] Howard finally pledged that a nationwide emissions-trading scheme would be in place in Australia by 2012. However, he still lost the general election in November 2007, largely because of his unpopular stance on climate change. One of the first acts of his successor, Kevin Rudd, was to ratify Kyoto.

European Union

The European Union is second only to the United States in its share of historical responsibility for the greenhouse problem.[24] However, its attitude has been very different. Rather than ducking the problem, the European Union has placed climate change squarely at the top of its agenda for most of the past decade.

The European Union committed itself as a bloc to a Kyoto target of reducing emissions by 8 percent from 1990 values by 2012—a target that looks very likely to be met,[25] though it will probably require some contribution from Clean Development Mechanism projects overseas. The European Union also pioneered a continent-wide carbon cap-and-trade scheme, which began in 2005 and which, despite the teething troubles we mentioned in chapter 10, is responsible for almost the entire global market in carbon emissions. And it has set out a comprehensive plan for how Europe will need to adapt to the changes that are already inevitable.[26]

Perhaps most important, the European Union's official policy is now to try to avoid the worst consequences of climate change by restricting temperature rises to no more than 3.5°F above the preindustrial level. As we showed in chapter 6, this might not now be achievable, even with a stabilization level of 450 ppm CO_2eq.

In May 2007 the Spring Council meeting of the European Union reaffirmed this stance, which had been set out in two earlier recommendation documents.[27] It called for developed countries to reduce their emissions levels by 30 percent above 1990 levels by 2020, aiming for a 60 to 80 percent cut in 2050. According to this plan, many developing countries would also need to cut their emissions in later decades, while being encouraged on to a low-carbon path through investments from the developed world.[28] Emissions levels would reach a peak in the next two decades before falling by 2050 to a global level 50 percent below where it is today.

Thus, the European Union has already made its negotiating position for a new global agreement very clear. It has also committed "without prejudice" to a unilateral continent-wide cut of at least 20 percent by 2020, explicitly to show leadership and to set the international ball rolling.

This has had an effect on wider groupings. The G8 summit in Germany in June 2007 issued an extraordinary statement jointly with the European Union and the five most rapidly developing nations that said: "We noted with concern the recent IPCC report and its findings. We are convinced that urgent and concerted action is needed and accept our responsibility to show leadership in tackling climate change. In setting a global goal for emissions reductions in the process we have agreed in Heiligendamm involving all major emitters, we will consider seriously the decisions made by the European Union, Canada and Japan which include at least a halving of global emissions by 2050."

This statement was criticized for its lack of binding targets, which had been summarily blocked by the United States. But it is the first document to which the world's top economies have put their signature that even mentions the need for a cut as large as this and on such a timescale. (Many commentators seemed not to notice that the desired target in this statement was for a reduction of *global* emissions by 50 percent by 2050, which, as we've already pointed out, would mean a cut of up to 80 percent for the G8 nations that signed.)

That there was any agreed statement on this most contentious of topics can be attributed to tireless lobbying from European G8 nations, especially the United Kingdom, the host nation Germany, and France.

France

Per capita: 9.0 tons/person
Historical: 6.6 tons/person
Total: 563 megatons
Change in emissions since 1990: −0.4 percent
Ratified Kyoto? Yes
Kyoto target for 2012: 0 percent

France shut down its last coal mine in 2004 and now generates some 80 percent of its electricity from nuclear power. While the nuclear option is anathema to many environmental groups, it has ensured that France has one of the lowest per-capita emissions rates in the industrialized world. Thanks to its nuclear plants, France is responsible for barely one-fifth of the average European emissions per unit of electricity generated, though the values from its transportation, industrial, and domestic heating are much higher than this.

The French government supported international efforts to curb emissions from early on in the negotiations. In 2002 the then president, Jacques Chirac, declared, "Our house is burning and we are looking the other way," and added that climate change was threatening "a planetary tragedy."[29] France suffered especially from the following year's heat wave, which did much to raise consciousness among its citizens of the dangers of climate change.

In March 2005 an Environmental Charter was incorporated into the French Constitution saying that "the choices designed to meet the needs of the present generation should not jeopardize the ability of future generations and other peoples to meet their own needs." That same year, an energy bill established the goal of decreasing energy use by 2 percent per year, continuing to support nuclear power, and putting particular effort into developing new, low-emissions energy sources.

In parallel with the EU proposals, the French energy bill also set out an ambitious long-term strategy to reduce the country's emissions below 1990 levels by 75 percent by 2050. Under the terms of the bill, the government is required to do its utmost to achieve this target, although it is not yet obligatory.

When the new president, Nicolas Sarkozy, was elected in 2007, he also upped the status of the environment in his government by creating a new "super-ministry" to cover the environment, energy, transportation, natural habitats, and land planning. Its head, Jean-Louis Borloo, is also Sarkozy's number-two man.

To achieve its cuts, France is focusing on new technologies, especially biofuels in the transportation sector, and strategies to make big improvements in efficiency. For instance, many efficiency improvements in individual homes—such as buying new, more efficient boilers—now earn considerable tax breaks. France also has relationships with many African countries involved in Clean Development Mechanism projects.

Germany

Per capita: 12.3 tons/person
Historical: 9.0 tons/person
Total: 1,015 megatons
Change in emissions since 1990: −17.5 percent
Ratified Kyoto? Yes
Kyoto target for 2012: −21.0 percent

Germany has a long and illustrious record of championing environmental causes in general, and climate change in particular. It was one of the first countries in Europe to make widespread deployments of wind and solar power. Thanks to its 1999 "100,000 roofs" program, which offered favorable loans to households installing solar panels, Germany now has the world's largest installed capacity of solar energy. It is also a world leader in manufacturing and exporting renewable energy devices. By one estimate, one in two turbines and one in three solar cells are made in Germany, and some economists have predicted that renewable energy could outstrip car manufacture as the country's most important export industry.

The impetus for domestic action slowed somewhat in recent years, in part because of economic and political problems associated with the reunification of the country. However, as chancellor, Angela Merkel has brought new energy to Germany's fight against climate change. A scientist by training, she understands the problem well and has made it one of the priorities of her administration.

One of the problems faced by Germany in tackling climate change—and one of the reasons for its relatively high per-capita emissions—is the country's extensive coal deposits. In spite of the wide take-up of "green" energy sources, half the country's electricity still comes from coal, which is the most polluting of all the fossil fuels (and much of that is now in the form of brown coal, which is less efficient and more polluting still). In March 2007 a pilot plant began operating in Brandenburg to attempt to capture carbon dioxide from a power plant and convert it to a storable form. So far, the attempt has been modest—the plant's capacity is a mere fifty megawatts and it's not yet clear where the captured CO_2 can be stored. But if it is successful, the Swedish company operating it plans to increase the scale.

Because of the economic downturn in the former East Germany after the fall of the Berlin Wall, Germany's emissions as a whole have fallen significantly since 1990, a fact that is reflected in their Kyoto target. But Merkel wants to go still further. In September 2007 she put forward a list of twenty-nine measures by which Germany could make deeper emissions cuts. These include all the usual suspects that we discussed in part 2. By putting them together, Merkel's government plans to reduce Germany's emissions by an extraordinary 40 percent by 2020—the world's most ambitious goal to date.

Merkel has also brought new energy to Germany's role at the forefront of international action. Like Tony Blair, Merkel made climate change one of the key pillars of the German G8 presidency, and it is largely thanks to the intense pressure applied by both these leaders and by President Sarkozy of France that the G8 + 5 meeting in Heiligendamm in June 2007 was able to issue its extraordinary joint statement on climate change.

Chancellor Merkel has also now formalized the Gleneagles Dialogue between the G8 and the five most rapidly developing nations, which will continue to try to establish the basis for an international agreement among the world's most polluting economies. This in turn could well be the basis for the fully global agreement on climate change that we urgently need.

United Kingdom

Per capita: 11.0 tons/person
Historical: 11.2 tons/person
Total: 656 megatons
Change in emissions since 1990: −14.2 percent
Ratified Kyoto? Yes
Kyoto target for 2012: −12.5 percent

Perhaps appropriately, since the Industrial Revolution started there, the United Kingdom has been one of the world's most proactive countries in both implementing national measures and driving the international process.

When the Labour Party came to power in 1997 they inherited a steep drop in greenhouse emissions since 1990, mainly because of the switching of much of the United Kingdom's power supply from oil to natural gas. Deputy Prime Minister John Prescott initially made climate change his mission and was one of the driving forces at the 1997 Kyoto meeting, which led to the Kyoto Protocol. In 1998 Chancellor of the Exchequer Gordon Brown imposed the Climate Change Levy on CO_2-producing energy sources (and also, curiously, on nuclear power). Much of the proceeds from this levy went to setting up and running the Carbon Trust and the Energy Saving Trust, two bodies set at arm's length from government, with the mission to raise the public profile of climate change and to work with businesses on reducing emissions by improving their efficiency.

In April 2002 the government introduced a scheme whereby utility companies were obliged to generate a certain proportion of their electricity from renewable resources (10 percent by 2010 and 20 percent by 2020, not including nuclear power), or to trade their way out of trouble by buying permits from other companies.

The United Kingdom also set up the world's first national carbon-trading scheme in 2004, which has since been folded into the European scheme. Being first has its advantages. The London Stock Exchange has certainly captured the carbon trade market, vying for global prominence only with Chicago's Climate Exchange, which has been operating since 2003 in spite of the lack of a nationwide trading scheme in the United States.

Since 1997 the United Kingdom's overall greenhouse emissions have continued to fall, mainly because of serious efforts to cut the emissions of methane from landfills, which have reduced the emissions of this powerful greenhouse gas by 55 percent. The nation will therefore have no problem meeting its Kyoto obligations.

In 2003 the United Kingdom unilaterally declared that it would reduce its carbon dioxide emissions by 60 percent by 2050, and challenged other countries to do the same. A white paper published that year provided a rough sketch of how this could be achieved, and set the interim target of reducing emissions by 20 percent by 2010, a level more stringent than the Kyoto one.

However, carbon dioxide emissions have been creeping up in spite of the above national measures, mainly because of the gradual retiring of Britain's aged nuclear power plants. There are also disadvantages in being first. The United Kingdom built the world's first commercial nuclear power station, some sixty years ago. But its current generation of power plants is now coming to the end of its life. Fifteen years ago 30 percent of the United Kingdom's electricity came from nuclear power. Today that figure is 19 percent and, unless new plants are built, by 2020 it will be down to 7 or 8 percent. Every time a nuclear power plant is replaced by one driven by fossil fuels, the United Kingdom's emissions rise inexorably. It is now highly unlikely that the nation will reach its own 2010 target.

Thus, in 2007 the government published a new white paper, giving much more detail on how to reach the 2050 target using a slew of energy-efficiency measures as well as bringing in new renewable technologies.[30]

To address the gap in current low-carbon technologies, one of us (David King), working closely with the (then) chancellor of the Exchequer, Gordon Brown, brokered the establishment of a public–private partnership called the Energy Technologies Institute to develop new, low-carbon technologies. It will receive funding of £100 million (nearly $150 million) per year for ten years, half to come from industry and half from the government. On the board already are motor companies, utilities, and oil companies, and negotiations are currently under way to fill the remaining available seats.

It's disappointing, certainly, that the 2010 target is now unlikely to be met. However, the nation has learned its lesson. To ensure that it does better in the future, the United Kingdom has also taken steps to enshrine the reduction commitments in law with the Climate Change Bill, which commits the country to a 60 percent drop in carbon dioxide emissions by 2050 and an interim 26 to 32 percent drop by 2020. It is the first country in the world to make such a long-term target legally binding. (This target will almost certainly need to be increased, but it is still a very important step.) There will also be an independent Climate Change Committee, which will work at arm's length from the government after the model of the Bank of England and will be allowed to use a whole raft of measures to ensure that emissions fall as required.

On the international stage, the United Kingdom has been a crucial player, in many cases stepping into the vacuum of leadership left by America's having turned its back (temporarily, one hopes) on climate change. As well as its avid participation in the Kyoto process and its role in setting up the European Trading Scheme, the United Kingdom has taken considerable steps to unite the world under a climate change banner. In 2005 Tony Blair made climate change one of the two main pillars of the United Kingdom's G8 presidency. He invited the five most rapidly developing nations, as well as other major industrial emitters, to the G8 meeting in Gleneagles. The result was the beginning of the Gleneagles Dialogue, which was followed by a further meeting between these countries in Mexico specifically to try to establish the grounds for a new international agreement.

The process was formalized in Germany in 2007, and—for its capacity to put pressure on the United States as well as draw the key developing countries into the debate—looks likely to be a key step in breaking the international deadlock. The prime minister also put personal pressure on President Bush to agree to the joint G8 statement on climate change issued in Germany in 2007.

When we finally manage to get a good agreement in force, individual nations will then need to implement it. Each will choose its own way, but there will be some common factors. In particular, each nation will need to:

❏ Set in place adaptation strategies to cope with the changes in the decades to come (see chapter 4).

❏ Switch to existing low-carbon technologies and (especially in the case of richer nations with more established science bases) fund research and development for new technologies (see chapters 8 and 9).

❏ Increase efficiencies and reduce power demand (see chapter 7).

❏ Engage in a combination of motivating factors that could include tax incentives and penalties, nationwide or international emissions, and permit trading schemes (see chapter 10).

❏ Find a way to tap the rising tide of consumer desire for action. For instance, governments will need to empower consumers with knowledge about how best to act, by encouraging accurate labeling of products to reveal their associated carbon dioxide emissions throughout their entire production process and setting regulatory standards for offsetting schemes (see chapter 14).

As should be clear from this chapter and the previous one, governments will also need to maintain pressure on all participants to come on board. Somehow, the nations that are already convinced of the dangers of climate change will need to convey this to the rest. It's easy to forget how vulnerable we are. Those of us who live in the richest nations are like the airplane passengers up in first class with the flat beds and the complimentary champagne. As climate change will come first to the world's most vulnerable citizens, the rest of us might lie back, relax, and think that there's no need to worry. But if the wings of the plane fall off, we all end up just as dead.

14

HOW YOU CAN CHANGE THE WORLD

It's easy to believe that global warming is somebody else's problem—other people will suffer and other people will come up with the solution. However, this is far from the truth. There's a clue in the name: "Global warming" is a truly global problem. None of us is safe from its effects (although some of us have a better chance of adapting to them). We are all part of the problem, and each of us will need to be part of the solution.

Today there are more than six billion human beings on Earth. By the middle of the century there will be nine and a half billion. Even without the problem of climate change we would be finding our resources running thin. With it, the population boom becomes even graver.

Thinking this way presents the human race as one massive blob. But in fact it's as individuals that we live our lives and make our choices. Every time each of us switches on a light, reaches for something in a supermarket, gets into a car or bus, or chooses what clothes to buy or which movie to see, we have all made a difference to the way the economy works. Choices like these have driven the world's economies ever upward in the twentieth century. They have also led to spiraling greenhouse gas emissions. Now we will all have to adapt our choices to the new realities of the twenty-first century.

The previous few chapters have shown how much of the answer to the problem of climate change will need to come from the top down, from governments, industries, and big international agreements. But none of this will ever happen unless we, the people, push from below. Through the choices we make about more or less every aspect of our personal lives, we need to drive producers and manufacturers along sustainable pathways. Through the choices we make in the ballot box, and through the pressure we put on our local representatives and communities, we need to drive politicians to set our countries on the right paths. We are the ones who have the power to change.

In chapters 7 ("More from Less") and 8 ("Planes, Trains, and Automobiles"), we pointed out many ways in which each of us can begin to tackle climate change in our own lives. Below, we also describe how to put pressure on politicians to ensure that they set the right national goals and reach the right international agreements.

Some of these points might seem familiar, but others may surprise you. And together these choices add up to something momentous. They will bring about nothing less than a global change in culture. As we've already mentioned, until now we have all treated energy as something that is almost free and infinitely available. We have wasted resources and ignored the perils of digging ever deeper into our planet's capital. To pull ourselves out of this mess we all need to change our attitude and realize how precious our energy resources really are.

Before we launch into specifics, here are a few words on two particularly contentious issues: carbon offsetting and food miles.

Going Carbon Neutral

There is much talk lately of "going carbon neutral." Since it's currently impossible to exist in the developed world without causing at least some greenhouse emissions, the only way to do this is to dip into the controversial waters of carbon offsetting.

. In principle, this is a good idea, akin to the Clean Development Mechanism of the Kyoto Protocol that we described in chapter 10. Simply put, when you buy a carbon offset you're compensating for, say, one ton of your own emissions by paying someone else to reduce theirs by the same amount. Since the air doesn't care where the greenhouse gas comes from, the result is a balance. You've "offset" your own emissions.

Usually, the "somewhere else" is a developing country, because that's where the cheapest and easiest opportunities are to make a difference. For instance, if a hospital in India is using kerosene for lighting, your offsetting money can fund the switch to solar panels. This could help India to leapfrog to the latest, cleanest technologies without having to pass through the various intermediate (and highly polluting) steps that Western countries did.

Hard-line environmentalists object to this idea in principle. They complain that offsetting lets people off the hook—and even discourages them from changing their own ways. Environmental journalist George Monbiot has, very wittily, compared carbon offsets to the medieval practice of the Catholic Church selling indulgences—pieces of paper saying that a particular sin was forgiven so that you could commit it without feeling guilty.

There is some strength in this argument, and carbon offsetting certainly shouldn't be your first recourse. There are plenty of other ways to try to cut down your emissions, including many that we list later in this chapter. Friends of the Earth, Greenpeace, and the WWF put out a joint statement about carbon offsetting in which they say that "we would encourage individuals, business and governments to first do all they can to cut down or avoid emissions of greenhouse gases before considering the purchase of offsets."[1]

But there are some levels beyond which you can't drop, and some forms of emitting (such as long-haul flying) for which there are no alternatives apart from not making the journey in the first place. In these cases, and as long as you're also making every effort to cut down in other ways, it makes sense to consider offsetting your emissions.

The next problem with offsets, however, is being sure that they actually do what they say. Offsetting companies are mushrooming, and there is currently little or no regulation. Your nice certificate saying that you have offset ten tons of greenhouse gases might not be worth the paper it's written on.

The problems here are the same ones that we mention in chapter 10 for the Clean Development Mechanism. How can you be sure that the offset you're buying (1) will really happen; (2) will really take up as much greenhouse gas as you think; and (3) wouldn't have happened anyway, even without your money?

For instance, in the early days plenty of companies offered to plant trees on your behalf. But trees take a long time to soak up carbon dioxide, and if they are burned either deliberately or accidentally during that period, the CO_2 goes right back into the air. Also, it's tremendously hard to calculate how much carbon dioxide a tree will take up during its lifetime. Another problem is that, away from the tropics, planting trees where there were none before can actually make things worse by changing the reflectivity of the land and encouraging it to soak up more sunlight.[2] And there's also the danger that people can claim carbon credit for plantations that they were planning to grow anyway.[3]

Without some kind of reliable benchmark, the carbon offsetting industry will always be vulnerable to deliberate fraud, but even well-meaning attempts to provide reliable emissions cuts can fall flat.[4] Various nongovernmental organizations (NGOs) have therefore clubbed together to create an independent standard for offsetting projects—the Gold Standard.[5] This is awarded only to renewable energy and efficiency projects, specifically because these encourage changes in behavior rather than just allowing the world to continue its reliance on fossil fuels. The projects are also rigorously and conservatively tested for whether they are truly additional (i.e., whether they wouldn't have happened anyway) and whether they will give rise to the reduction in emissions they claim.

Thus, if you do decide to offset any or all of your emissions, we recommend that you go only for projects that bear this standard. Anything else might turn out to be just snake oil.

Food Miles

As climate concepts go, this one seems to be as simple as it gets. Food has to travel a certain distance from farm to plate. The farther it goes, the more transportation emissions it causes. Therefore, it should be best, from a climate perspective, to buy food that has come the shortest distance and hence has the lowest associated "food miles."

However, the true picture is more complex. For one thing, it makes a big difference how the food has traveled. Planes are worst of all in emissions terms, followed by cars and heavy goods vehicles (HGVs), though if HGVs are fully loaded they can be relatively efficient, emissions-wise.[6]

More surprisingly, even when you've factored in the mode of transportation, the simple idea of food miles can lead you in the wrong direction. It's important to look at the entire greenhouse emissions a particular foodstuff has caused, and not just on the road. For instance, growing tomatoes in greenhouses in the United Kingdom, with all the heating and lighting that takes, can cause more overall emissions than growing them in sunny Spain and then transporting them to Britain.[7] And a study in

2006 by the Agribusiness and Economics Research Unit at Lincoln University in Christchurch, New Zealand, concluded that dairy products made in New Zealand and then exported to the European Union were twice as energy efficient as those made locally, and sheep meat was four times as efficient.[8]

Partly, such anomalies reflect our increasing desire for food that's out of season locally or could never be grown in the local climate. But they serve to illustrate that looking at how far something has traveled doesn't necessarily help you decide where the lowest emissions lie. Confusions like this would be solved if national governments insisted that products be fully labeled with their cradle-to-grave emissions, which is a policy we should all be pushing for.

In the meantime, a good rule of thumb is that if something is in season and grown locally, it's likely to have produced fewer emissions than something that's out of season and has come from far away. If a food is out of season, on the other hand, it's harder to tell.

But there's another sting in the food miles tail: the issue of sustainable development. Avoiding food that has been airfreighted in from developing countries could have a seriously detrimental effect on their economy, without necessarily doing a lot to help climate change. For instance, the International Institute for Environment and Development published a report in 2006 saying that the importation of fruit, vegetables, and flowers into the United Kingdom from sub-Saharan Africa was a vital part of the economy of the countries involved, and supported up to one and a half million livelihoods in some of the poorest countries on Earth. The report also points out that, even though most of the produce is airfreighted, because of lack of infrastructure in the home countries and the danger of perishing en route, the imports still contribute less than 0.1 percent to the importing country's overall greenhouse emissions. Emissions from buildings, personal transportation, agriculture, and even road miles for transporting food within the country itself all contribute much more to the problem.[9]

Calculate Your Carbon Footprint

A good start when setting out to reduce your own carbon emissions is to find out what they are. There are now many calculators available on the Internet that will help you determine the amount of greenhouse gas emissions for which you are personally responsible. Until now we have all been suffering from a climate version of attention deficit disorder, or perhaps it should be called information deficit disorder. Most of us have no idea how much greenhouse gas we are responsible for, or even which areas of our life produce the most or are most amenable to change.

Stay Informed

The climate story—scientific, technological, and especially political—is developing every day. It can be hard to see your way through the blizzard of information, but it does pay to stay informed. At the back of this book we list some good Web sites, and trustworthy sources of information. (As with everything else in life, when you're deciding who to trust, one of the best ways is to look at what they have to gain.) We'll also do our best to keep our own Web site, www.thehottopic.net, abreast of the latest developments.

Be Open-Minded

You can make a big difference by being prepared to consider some of the more controversial solutions to the problem. For instance, the siting of new wind farms is often bogged down in local planning permissions, and nuclear power has a vocal set of opponents who often aren't prepared to consider the pros as well as the cons. Keep your mind as open as you can; consider all the arguments in light of the serious threat of climate change. And remember that any kind of change—especially a shift in culture of the magnitude we need—also requires difficult choices. We human beings are deeply conservative in principle, but in practice we have used our flexibility and ingenuity to great effect in the world. These are the qualities that we all need to use for the good of the climate.

It will also be important to be receptive to imaginative new ideas to help us tackle the problem. One intriguing suggestion is that of personal carbon allowances, often advanced in the form of a "carbon credit card." This would work like a prepay card for a mobile phone. Each adult would receive an allowance of carbon emissions to be used against, say, electricity, heating, and personal transportation. Every purchase in one of these areas would require a swipe of the credit card. If you need more than your allowance, you pay to top your card back up. If you need less, you can sell your surplus for money. The system for paying extra or receiving money could be as simple as a bank cash machine.

Most versions of this idea assume it will apply only within one nation. But there are some interesting potential variations. For instance, we could each have a personal allowance for our domestic emissions, but any international travel would need carbon credits from a Gold Standard–accredited project overseas.

One major advantage of an approach like this is that it would focus our attention on which emissions we are responsible for and encourage us to do something about them. The British government commissioned a preliminary report to study the scope of the various schemes proposed to date, and it makes fascinating reading.[10]

Vote with Your Ballot, Your Wallet, and Your Feet

When you vote, check where the candidates stand on climate issues and put pressure on them to take global warming seriously. There are plenty of Web sites that will tell you what your representative has said on the subject, and we list some of these at the end of the book. Even if an election is a long way off, lobby your local representatives by e-mail or letter. Chapter 11 explains what sort of international climate agreement they should be advocating. Tell them they need to seek a global accord to keep greenhouse gases below 450 ppm CO_2eq. Ask them to push for important national policies, especially labeling all products according to their greenhouse impacts so that you, the consumer, can make informed choices.

As we've outlined above, even without full information about cradle-to-grave emissions, you can still make a difference by what you buy and how you travel. It's worth keeping the climate in mind with every financial decision you make.

Act Local to Change Global

The evidence is growing around the world that pressure from local organizations, city councils, and states can sway national policies on climate change. We have already described in chapter 13 how action in the United States by local city mayors, and by California and the northeastern states, has helped cast President Bush's stance against action in an increasingly untenable light. Then there's the city of Austin, Texas, which lies in the heart of oil country in one of the most conservative states in the union but is one of the most progressive when it comes to dealing with climate change. Their Austin Climate Protection Plan is so comprehensive it should shame any government into action.[11] They plan to make every aspect of the city's operations carbon neutral by 2020, to implement "the most aggressive utility greenhouse gas reduction plan in the nation," make Austin's building codes the most energy efficient in the United States, and encourage all businesses and individuals to reduce their carbon footprint to zero.

In fact, the world over, individual cities, towns, and communities are not waiting for governments to act; instead, they are taking the issue of climate change into their own hands. In the United Kingdom, for example, the city of Newcastle has ambitions to be the first in the world to go fully carbon neutral. It has put in place a host of policies, starting with the local council itself, but then reaching farther to encourage both local businesses and individual citizens to change their climate ways without harming their lifestyles. The initiative is hugely popular, and now local politicians there have to vie with one another to show their commitment to tackling climate change.[12] And while Canada's national government continues to be recalcitrant about climate change, most recently in the United Nations conference in Bali in 2007, it is facing a barrage of pressure from opposition parties and environmental groups who have issued a series of legal challenges aimed at forcing the government to meet its Kyoto obligations. Meanwhile, the province of Quebec and the city of Toronto have each set out unilateral programs for reducing greenhouse emissions, including ambitious targets for the next decade and beyond. Quebec has even imposed a "hydrocarbon levy" on energy from fossil fuels. Similar pressure on Australia's national government by individual states helped turn a national emissions-trading scheme from distant dream to imminent reality. (It also ultimately cost former prime minister John Howard his job, which is something that other politicians would do well to note.)

Pressure like this isn't confined to the developed world. Western Cape province in South Africa is likely to suffer greatly from future changes in climate, and the inhabitants know it. And so, instead of sitting back and waiting for the national government to act, the province's regional government has decided to take matters into its own hands. In July 2007 it published a detailed strategy document explaining how it plans to improve efficiency, switch to renewable energy sources, and reduce the overall carbon footprint.[13] The targets are impressive. By 2014 the regional government plans to increase overall energy efficiency for the region by 15 percent, generate 15 percent of its electricity from renewable resources, and reduce its overall greenhouse emissions by 10 percent. This is extraordinary news. Such a unilateral step, taken in a developing country, could have political repercussions that matter even more than the actual emissions cuts.

The bottom line of all this is that pushing to change the ways of your school, workplace, or local council really can make a difference. As local initiatives spread and as local politicians get the message that climate policies matter, small seeds can flourish quickly and very effectively. If Newcastle, Austin, and the Western Cape province can do it, so can we all.

Be Positive

Don't be "greener than thou." The evidence suggests that making people feel guilty makes them less likely to act, not more. And none of us has the right to be smug, no matter what we are doing to reduce our emissions. Since the day we were born, each of us has been inadvertently contributing to the greenhouse problem, and the wealth we now have has been built on the emissions of previous generations. Anything we do now is still doing no more than paying back a fraction of that old debt. Thinking otherwise is to believe that, once you've bought your Prius, climate change is now somebody else's problem. It's not. We're all responsible, and will continue to be until we've stopped global greenhouse emissions in their tracks.

It is, however, a good idea to admire other people's efforts to cut back. The latest research suggests that people are more likely to act altruistically over climate if they feel it enhances their reputation.[14] Gentle persuasion can also help, as can taking the lead. If you've implemented ways of cutting carbon at home, see if you can set up the same ways at your place of work or school.

Opinion polls differ on how people feel about climate change and what they currently believe. But there is still a substantial proportion of people who think the problem isn't important, or that it doesn't apply to them.[15] One way to help is to talk the people you meet into taking climate seriously. (Our handy guide to common climate myths should help here.)

It's also important not to fall into the trap of thinking that what you do to change your own life doesn't make a difference. It does. Even if the actual amount of emissions that you've cut is only a tiny proportion of the global total, the only way we will solve the climate problem is to do so together. UK negotiators have already discovered that the country's unilateral decision to cut its emissions by 60 percent by 2050 meant that developing countries such as China and Brazil immediately took them seriously. And the actions of individuals really can put pressure on the government to respond, as the situation in the United States clearly shows.

Above all, don't despair. The climate problem is certainly a hard one, but it's not intractable. If you still need convincing, here's a final image that should cheer you up. When we learn to capture sunlight efficiently, we'll have all the energy we need on our very doorstep. Enough energy to power the entire world's requirements can be collected from the land area contained in these six tiny squares.[16]

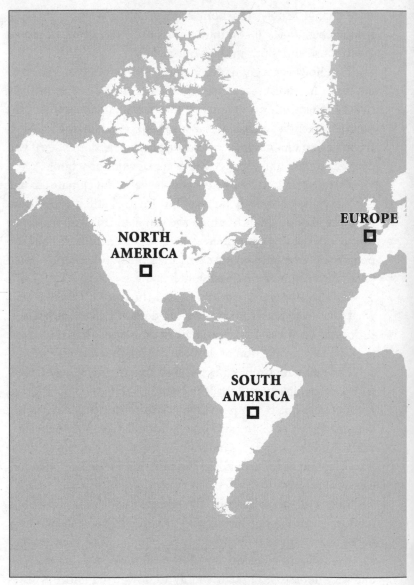

If all the sunlight falling on each of the areas inside these six squares could be completely harvested, that would supply the entire world's energy needs. (Source: Nathan Lewis)

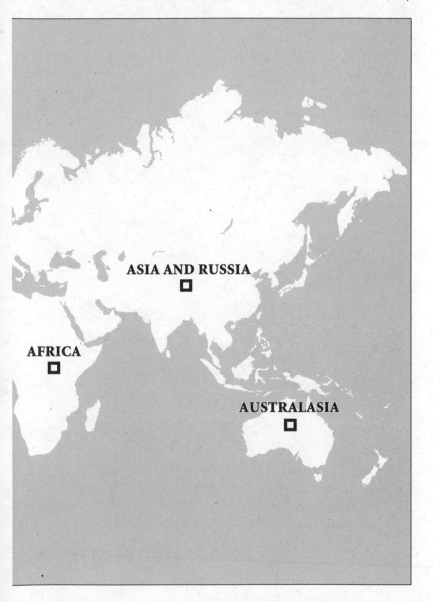

ACKNOWLEDGMENTS

The authors would like to thank the following people:

For inspiration and advice, especially on the subject of climate change:
Richard Alley, Michael Bender, Henry Derwent, Laura Garwin, John Houghton, Geoff Jenkins, Ralph Keeling, Oliver Morton, Robert Napier, Fred Pearce, John Pyle, Dan Schrag, David Warrilow, and Jeremy Webb.

For reading and commenting on all or part of the manuscript:
Fred Barron, David Bodanis, Partha Dasgupta, Michael Evans, Tim Flannery, Jim Lovelock, Rosa Malloy, Oliver Morton, Nick Rowley, and Bob Watson.

For support, especially during writing:
Fred Barron, Stephen Battersby, Yann Golzio, Ben King, Toby King, Emily King and Zach King, Jane Lichtenstein, Rosa Malloy, Damian Malloy, Dominick McIntyre, Philippe Mottin, Christine Nesme, Jean-Chris Sinardet, Helene Sinardet, Sammy and Minnie Sinardet, Simon and Anita Singh, Karen Southwell, Helen, Ed, and Christian Southworth, Jayne Thomas, and John Vandecar.

Special thanks to our editors Bill Swainson and Emily Sweet at Bloomsbury and Andrea Schulz at Harcourt, and to our agents Michael Carlisle and Susan Hobson at Inkwell. And extra special thanks to Michael Evans and Hubert Sinardet, without whom this book could not have been written.

APPENDIX

Climate Myths, Half-Truths, and Misconceptions

The Science of Warming

It's not really warming.

Yes, it is. The years 1998 and 2005 were the joint hottest on record (that is, in the past 150 years). The years 2002, 2003, and 2004 were, respectively, the third, fourth, and fifth warmest on record. In fact, ten of the past eleven years have been in the top eleven. Reconstructions of past temperatures using corals, ice cores, and other techniques show that the temperature is hotter now than it has been for at least one thousand years and probably longer (see chapter 1).

It was warmer during the Middle Ages than it is now.

No, it wasn't. Temperatures are higher now than they have been for at least one thousand years (see chapter 1).

The ice cores show that temperature goes up before carbon dioxide at the end of ice ages, so CO_2 can't cause warming.

The ice cores do show a time lag between the onset of warming and the rise of carbon dioxide at the end of each ice age. However, that doesn't mean that carbon dioxide doesn't cause warming. In fact, nobody thinks that carbon dioxide is what causes the ice ages to stop. Instead, a wobble in our planet's orbit changes the distribution of the sunlight we receive. That causes a little warming, which sets in motion other processes to release carbon dioxide. For instance, a warmer ocean holds less carbon dioxide, so more makes it into the atmosphere. Also, a warmer, wetter climate means that less iron-rich dust is blown out onto the sea. In ice ages, this iron probably feeds the growth of plankton that soak up carbon dioxide. Cut off this nutrient source and you have fewer plankton, and more CO_2.

The important part is what happens next. That extra carbon dioxide causes warming in its own right, which begets more carbon dioxide, which begets more warming. According to the ice cores, the overall warming takes some five thousand years to complete. The slight shift in sunlight from Earth's wobble wouldn't be nearly enough to do the job by itself. Instead, most of the warming (after the first few hundred years) comes from feedbacks, including carbon dioxide.

That's a perfectly natural process and is used by climate scientists only to show that carbon dioxide does in fact warm the planet if it goes up by even a small amount. The increase at the end of an ice age, by the way, took carbon dioxide levels much lower than they are today (see chapter 1).

Temperatures were going down in the middle of the twentieth century even though carbon dioxide was rising, so CO₂ can't cause warming.

This is a very popular argument put forward by climate skeptics, but it skates over the fact that scientists can explain why temperatures dropped a little in the middle of the twentieth century in the northern hemisphere. In fact, burning dirty fossil fuels produces sulfur-containing particles called aerosols, which reflect sunlight back to space and thus help to cool the planet.

They thus act in a sort of tug-of-war with carbon dioxide, creating cooling that offsets the greenhouse warming.

Researchers now believe that these aerosols were responsible for the slight cooling that took place between about 1940 and the late 1960s. The reason they didn't keep cooling is twofold: We cleaned up our act by banning the fuels that were choking our cities, and carbon dioxide levels rose so high that they won the tug-of-war. Also, the reason this cooling was seen only in the northern hemisphere is that there was neither enough land mass nor enough industry in the southern hemisphere to produce enough aerosols to counteract the effect of carbon dioxide (see chapter 2).

Human carbon dioxide emissions are tiny compared with natural sources, so they can't be important.

It's true that natural sources give off much more carbon dioxide than humans do, but it's also true that natural sources take up much more carbon dioxide, too. Roughly speaking, the natural world is in carbon balance. Our own human emissions are what have thrown the world off this balance.

Carbon dioxide concentrations are tiny compared with the other gases in the atmosphere, so they can't be important.

Carbon dioxide does indeed make up only a tiny part of the atmosphere; there are just four molecules of CO_2 in every ten thousand that the air holds. But carbon dioxide also punches considerably above its weight in the greenhouse effect. The main constituents of air—nitrogen and oxygen—can't trap infrared radiation at all. By contrast, carbon dioxide, even in small quantities, is a very effective trapper of heat. It also triggers feedbacks that enable the air to soak up more water, which is a greenhouse gas in its own right. Thus, changing CO_2 and the other greenhouse gases by even a small amount can affect the temperature of the entire atmosphere, just as adding a few drops of ink can change the color of a bath full of water (see chapter 1).

Humans aren't responsible for the warming—it's all been caused by a natural cycle/changes in the sun.

In fact, the warming over the past few decades has human fingerprints all over it. There's no natural cycle that can explain what we've seen, and the sun has been going in the wrong direction. (Left to itself, the sun would have caused a slight cooling!) What's more, models tell us that the increased greenhouse gas concentrations that we have had should cause exactly the changes that we have already seen. There's no more room for doubt in this. Wedunnit (see chapter 2).

Impacts of Climate Change

Earth's climate is constantly changing, so why should we worry?

Our planet's climate is certainly very restless, and in times past it has been both much warmer and much colder than it is today. But the whole of human civilization has been built around the climate we have now. For instance, just a few hundred thousand years ago, sea level was several feet higher than it is now, but back then there were no coastal cities waiting to be drowned (see chapter 1).

A little warming could be a good thing.

There are a few parts of the world where a little warming might indeed be a good thing. For instance, in the middle latitudes—northern Europe, the United States, and parts of Russia—a temperature rise of up to two degrees should actually increase overall crop yields. However, the problem with climate change is that you can't pick and choose its effects. Even for these countries, the increased food production will also come with northward movement of the storm tracks (less rain, more fires and droughts), more intense individual rainstorms (risk of flooding), rising sea level, more intense storms coming in from the oceans (threats to coastal cities), and more intense heat waves (mass mortality and threat to crops). In the poorest parts of the world, the effects will be more severe with even a modest rise in temperature. In fact, that's already happening. Even if you didn't want to act on this for the sake of social justice, you'd need to realize that in a fully global economy, what affects some countries will eventually hurt the rest.

The chances are that we are already going to experience the effects of a few degrees' warming in any case—that much is too late to stop. But action that we take now could help us avert worse warming, in which even the richest, most northerly countries would suffer (see chapters 4 through 6).

If the world warms, fewer people will die from the cold.

It's true that there will be fewer extreme cold events in the world, and the reduction in deaths from cold will almost certainly outweigh the increase in deaths from heat waves. But that doesn't mean humans will survive better overall. As well as direct deaths from the heat, people will also be more vulnerable to flooding, infectious diseases, and starvation. The IPCC report is unequivocal about this. "Impacts and Adaptations," Working Group II, chapter 8, says clearly that in terms of numbers of lives these increased dangers far outweigh the reduction in deaths from the cold.

Moreover, many of these deaths are also preventable as long as we adapt to the changes that are already inevitable (see chapter 3) and reduce greenhouse emissions to stop the changes getting worse (see chapter 6 and following). As German scientist John Schellnhuber put it, we need to "avoid the uncontrollable and control the unavoidable."

"Climate Porn," or Disasters That Aren't Necessarily Waiting to Happen

The ocean conveyor belt is about to switch off and plunge the world into an ice age.

In fact, though it was a popular disaster scenario in the Hollywood movie *The Day After Tomorrow,* this would be very unlikely to happen, at least in this century, even if we didn't act to curb greenhouse gases (see chapter 5).

Antarctica is about to slide into the sea.

The old, cold eastern part of Antarctica is very unlikely to melt even if temperatures were to rise dramatically. Nobody knows for certain exactly how safe the more vulnerable West Antarctic Ice Sheet is, but we can probably save most of it from melting if we act quickly (see chapter 5).

The Amazon rain forest is already doomed.

Though one model suggests that large parts of the Amazon could be vulnerable by the middle of the century, most say that the Amazon would be likely to stay intact until the end of the century even if we did nothing to combat greenhouse emissions. If we switch off the chain saws and divert our attention to cutting down emissions instead, the Amazon should be around for a very long time to come.

Misattributions and Misconceptions

The disappearing snows of Kilimanjaro prove/ disprove the existence of global warming.

Mount Kilimanjaro is a special case that tells us absolutely noth-ing about global warming. Its ice cap has certainly existed for some eleven thousand years, is currently retreating, and may well soon disappear. But we don't yet know exactly why. Because the retreat began in the early nineteenth century, those initial movements can't have been triggered by human-induced cli-mate change (although they might have come from some other human trigger, such as land-use changes). However, global warming might now be helping to tip it over the edge. The fact that the glacier has survived intact through thousands of years of natural climate fluctuations, and is only now on the point of vanishing completely, supports this idea. However, researchers have only a few years of detailed measurements on Kilimanjaro's glacier, and they won't be sure what is causing it to vanish until they have many more years of data.

All the fuss about Kilimanjaro is really just a distraction from the bigger picture. Glaciers are retreating the world over, from tropics to poles, and in many of these cases global warm-ing is clearly to blame.

The war in Darfur/Hurricane Katrina was caused by global warming.

In fact, we don't know. Scientists can't reliably attribute any single event to global warming. However, they can say that more strong hurricanes/droughts/floods/famines are on their way as a result of climate change. We can also learn from our past troubles. Whether or not the events in New Orleans and Darfur were "caused" by global warming, they still show us how devastating climate changes can be if we are not properly prepared (see chapter 3).

The Power of Technology

Nothing bad will happen for decades, so we should just wait for new technology to come along and solve the problem.

We will need new technology to help us out of our climate crisis. But as well as developing this, we will need to start using the technologies and energy-efficiency strategies we already have (see chapters 7 through 9). Dangerous climate change is with us, and there will be more to come in the next few decades whatever we do today (see chapters 3 and 4). Even more important, the longer we wait, the harder it will be to avert the worse consequences of future climate change. As we explained in chapter 6, we need to keep greenhouse gas levels below 450 ppm CO_2eq. To achieve that, emissions will have to peak within the next fifteen years and then start to fall. The only option is to act now.

Economics

We can't afford to tackle climate change.

In fact, tackling climate change turns out to be surprisingly cheap. Many of the strategies we need to employ to increase efficiency will actually save money where it has previously been wasted (see chapter 10), and investment in new technologies could well lead to a growth spurt.

We'd be better off spending the money on aid.

This is at best misguided. For one thing, if the money is not used for dealing with climate change, it's very unlikely to go to aid instead. (For instance, the International Energy Agency predicts that more than $20 trillion will need to be invested between now and 2030 to meet the world's growing hunger for power. The choice is between investing this in fossil fuels or in low-carbon forms of energy, rather than what else to spend the money on.)

Also, this is not about trying to solve all the troubles in the world—which is just as well, since that goal has spectacularly evaded us so far. Instead, it's about solving a specific problem faced by all of us, so we can leave behind a climate that our grandchildren will be able to weather.

The next two decades offer our only possible window of opportunity to rein in greenhouse gas levels to one that will achieve this goal. After that, no future generations of humans would be able to keep the greenhouse problem in check. For us, or our children, or our grandchildren, it would be too late.

Because of this, we believe it's better to think of the costs of climate change as an insurance policy against a dangerous future rather than as a handout. And put that way, it becomes a bargain. The amount of money that we would need annually to deal with climate change is barely half the global insurance industry's annual turnover (see chapter 10).

Politics

There's no point in doing anything because China's new power stations will swamp the effects of everywhere else.

We will never come up with a global agreement on climate change until everybody takes responsibility for their emissions. Yes, China is the biggest overall emitter, and one of the fastest growing. But it still puts out much less greenhouse gas per capita than the entire developed world. It has also been responsible for very little of the problem to date. To have any chance of persuading developing countries to join an international agreement, the developed countries—to date the main cause of the problem—will need to tackle their own emissions. UK negotiators found that their government's unilateral commitment to reducing the nation's emissions by 60 percent by 2050 made a big difference in how they were received by the developing world, even though the actual amount involved is only a small proportion of the global total (see chapters 12 and 13).

It's already too late, and there's nothing we can do.

It *isn't* too late, and there *is* something we can do. Though we can't stop the initial effects of climate change, some of which are already here, we do still have a chance to hold off the very worst things the climate could throw at us. In this book, we have laid out the solutions to the problem: adaptational (chapter 3), technological (chapters 6–9), economic (chapter 10), and political (chapters 11–13). We have also shown what you can do to make all this happen (chapter 14). This is neither the time to panic nor to bury your head in the sand. It's the time for action.

SELECTED GLOSSARY

"Business as usual" This phrase is used throughout the IPCC reports and in many other greenhouse analyses to refer to a projected scenario where nothing is done to curb greenhouse emissions in the future.

Carbon Capture and Storage (CCS) The idea is to burn fossil fuels in the normal way to generate power, but then to grab the carbon dioxide and bury it before it can escape into the atmosphere. This has the great advantage that it can remove emissions from traditional fossil-fuel plants, thus buying the world some time to develop new low-carbon alternatives. CCS is likely to be especially important for countries like India and China, which are currently exploiting their vast coal reserves at an increasing rate to fuel extremely rapid economic growth. However, it is not yet at a commercial stage, and plants using CCS will always be more expensive than those that simply vent their emissions into the atmosphere, so some kind of economic mechanism will be needed to make it viable.

Clean Development Mechanism (CDM) A mechanism devised for the Kyoto Protocol in which a (rich) developed country can gain carbon credit by setting up a program to reduce emissions in a (poorer) developing country, such as India.

CO_2eq This is shorthand for "carbon dioxide equivalent" and is a measure of the combined effects of all greenhouse gases: carbon dioxide, nitrous oxide, methane, and the rarer trace greenhouse gases such as chlorofluorocarbons.

It's an important number because different greenhouse gases are more or less effective at causing warming. So to find out how much warming we can expect from all of them, it's not enough simply to add their concentrations together. Instead, researchers give a different weight to the concentration of each gas according to its greenhouse potency, using carbon dioxide as the standard.

The resulting number gives the value of pure carbon dioxide that would have the same warming effect as all the greenhouse gases combined. It is a much better measure of the greenhouse effect than simply taking carbon dioxide on its own, and is the one we have used throughout this book.

The Intergovernmental Panel on Climate Change (IPCC) is an international body formed by scientists and government representatives to assess the risk posed by human-induced changes in climate. It was created in 1988 and is open to all members of the United Nations and the World Meteorological Organization. The IPCC doesn't carry out its own research. Instead, it produces reports that synthesize the current state of climate science.

Because the IPCC must reflect the consensus view of all its many contributors, it has a reputation for being conservative. It is also widely considered to be the definitive authority on the science of climate change. In 2007 it won the Nobel Peace Prize.

In this book, when we write "the IPCC report" we are referring to the latest one, the "Fourth Assessment Report," which was published in November 2007 by Cambridge University Press. The report comprises four volumes: the scientific basis, impacts and adaptations, mitigation of climate change, and an overall synthesis report. It runs to several thousand pages, has taken six years to complete, and contains contributions from thousands of scientists and government representatives from more than 130 countries.

Joint Implementation (JI) A mechanism devised for the Kyoto Protocol, in which a (rich) developed country can gain carbon credit by setting up a program to reduce emissions in a (poorer) relatively well-developed country, such as one from the former eastern bloc.

Kyoto Protocol The Kyoto Protocol was born out of the UN-FCCC (see below) in 1997. It sets out emissions targets for all signatories, with the aim of reducing global greenhouse emissions by 5 percent by 2012. For more explanation of the strengths and weaknesses of the Protocol and the need for a new international treaty, see part 3, especially chapters 10 and 11.

ppm This stands for "parts per million" and is the usual measuring unit applied to greenhouse gases, since they exist in relatively small quantities in the atmosphere. One ppm is 0.0001 percent.

United Nations Framework Convention on Climate Change (UNFCCC) This is the world's first international climate treaty. It came into force in 1994 and has since been signed by 189 countries—including the United States, the one nation that refused to sign the Kyoto Protocol. The stated objective of the treaty is to achieve "stabilization of greenhouse gas concentrations in the atmosphere at a level that would prevent dangerous [human-induced] interference with the climate system."

NOTES

Preface

1. D. A. King, *Science*, vol. 303, pp. 176–7, 2004. This article was based on the Zuckerman Lecture, "The Science of Climate Change: Adapt, Mitigate or Ignore?" given by David King in 2002 and available on the web at www.foundation.org.uk/events/pdf/20021031_King.pdf.

Chapter 1: Warming World

1. This is dangerous, because even something that seems like a concrete measurement of temperature can turn out to be nothing of the kind. A case in point is the famous "frost fairs" held on a frozen Thames River in London. Although certain freezing events did take place in the period known as the Little Ice Age, the river has actually frozen over twenty-two times since the thirteenth century, including during some periods that were warm elsewhere in the world. The real reason turns out to be local cold snaps, combined with narrow bridge spans that kept the surging tide safely downstream. No complete freezes have happened since 1835, when London Bridge's piers were widened, allowing the tide to flow much further upstream. See P. D. Jones, T. J. Osborn, and K. R. Briffa, "The Evolution of Climate over the Last Millennium," *Science*, vol. 292 (5517), pp. 662–67, April 27, 2001, DOI: 10.1126/science.1059126, and the discussion at http://www.realclimate.org/index.php/archives/2006/03/art-and-climate/.

2. It's not quite as simple as it sounds, because growth can be affected by amounts of sunlight, moisture, and even how much the tree grew the year before. But researchers have become expert at assembling all these factors to read the trees' record books.

3. Oxygen comes in different flavors called "isotopes," each of which has a slightly different atomic weight. Changing temperature alters the ratio in which these isotopes are incorporated from air into the snow.

4. For much more detail about this, see "Climate Change 2007: The Physical Science Basis" (written by Working Group I and subtitled "Working Group I Contribution to the Fourth Assessment Report of the IPCC Corporate Author Intergovernmental Panel on Climate Change," and hereafter referred to as IPCC WGI), chapter 6, section 6.6, IPCC reports (Cambridge University Press, 2007), especially figure 6.10c. Full details can be found at http://www.cambridge.org/browse/browse_highlights.asp?subjectid=710.

5. It's hard to tell, because there are fewer records from the southern hemisphere—thanks to the lower landmass as well as less overall research—so these conclusions are dominated by northern hemisphere temperatures. See Jones, Osborn, and Briffa, "The Evolution of Climate over the Last Millennium"; and Timothy Spall et al., "The Spatial Extent of 20th Century Warmth in the Context of the Past 1200 Years," *Science*, vol. 311, p. 841, 2006, DOI: 10.1126/science.1120514.

6. Quoted in Spencer Weart, *The Discovery of Global Warming* (Cambridge, Mass.: Harvard University Press, 2003), p. 1.

7. See IPCC WGI, chapter 6, especially Figure 6.10c, which shows an assembly of all the different reconstructions of temperature.

8. IPCC WGI, chapter 3.

9. The last ice age was between 3° and 5°C cooler than present; see IPCC WGI, chapter 6. Also note that the temperatures observed to date have been masked by two effects: the cooling effect of particles of air pollution and the time lag caused by the thermal inertia of the ocean.

10. See IPCC WGI, chapter 3. Nineteen ninety-six was the only recent year that stayed relatively cool.

11. IPCC WGI, Summary for Policymakers.

12. The most important greenhouse gas in the atmosphere (apart from water vapor) is carbon dioxide, but there are also substantial amounts of methane and nitrous oxide (otherwise known as laughing gas) and small amounts of a set of gases containing chlorine, which, as well as being greenhouse gases, also destroy the ozone layer.

13. Water in the atmosphere is extraordinarily dynamic. Between being sucked up from the ocean and rained out again via a cloud, an average water molecule will inhabit the air for just ten days.

14. This "water vapor feedback" formed the basis for one of the strongest critical arguments against the dangers of climate change. Respected

researcher and notable skeptic Richard Lindzen, a professor at MIT, believed that the upper part of the troposphere would dry out as carbon dioxide increased, and that this would balance the increased wetness lower down. However, measurements have since proved that argument incorrect. There's more about this in chapter 2.

15. We don't have space here to go into the history of how the greenhouse effect was discovered, though it's a fascinating story. For those interested, see Weart, *The Discovery of Global Warming,* or Gabrielle Walker, *An Ocean of Air* (London: Bloomsbury, 2007), chapter 3.

16. If air is in relatively short supply, some of the carbon will emerge in the form of black soot, which is why when you burn things that used to be alive—such as toast—they go black. Some will probably also be emitted as carbon monoxide, the deadly gas that comes out of car exhaust pipes and that people use to commit suicide.

17. For much more on this, see the excellent chapter 8, "Digging Up the Dead," in Tim Flannery, *The Weather Makers* (London: Penguin Books, 2007).

18. Although in principle the process of making food from carbon dioxide, eating it, and breathing the carbon dioxide back out ought to be carbon neutral for humans, agriculture has changed that. Though our hunter-gatherer ancestors were indeed in carbon harmony with their food, the energy we put into farming, and the greenhouse gases this emits, mean that breathing is no longer a carbon-neutral activity. See chapter 8 for more about the direct emissions from agriculture.

19. It might seem surprising that ice can be different ages in different parts of the continent. But the ice in Antarctica flows constantly, so although the continent has been ice-capped for millions of years, the bottommost ice at some of the thickest points has long since been squeezed out like toothpaste from a tube and disappeared toward the coast. In some places it has also melted, since the ice at the base of the cap is close to the melting point. The age of the ice that remains depends on how much snow falls in an average year. More snow falls at Vostok than at Dome C, so for the same thickness of ice sheet the ice at the base is younger.

20. See J-R. Petit et al., "Climate and Atmospheric History of the Past 420,000 Years from the Vostok Ice Core, Antarctica," *Nature,* vol. 399, pp. 429–36, 1999; and R. Spahni et al., "Atmospheric Methane

and Nitrous Oxide of the late Pleistocene from Antarctic Ice Cores," *Science*, vol. 310, pp. 1317–21, 2005. The ice cores also show very dramatically that higher carbon dioxide (and methane) always goes hand in hand with higher temperature. The two march in impressive lockstep. See the section "Climate Myths, Half-Truths, and Misconceptions" for the difference in timing between rising temperature and rising carbon dioxide, and why it isn't relevant to this argument.

21. The curve also shows the carbon dioxide levels snaking up and down a little each year as the air "breathes." Because there is more land in the northern hemisphere than the southern, each northern summer carbon dioxide levels drop as plants bind it up into food, and each northern winter levels fall again as the plants slow down or die off.

22. See IPCC WGI, chapter 6, especially Question 6.2.

23. There are also recent signs that methane may be leveling off, although nobody yet knows why, so it's not clear whether this good trend will continue. See I. J. Simpson, F. S. Rowland, S. Meinardi, and D. R. Blake, "Influence of Biomass Burning During Recent Fluctuations in the Slow Growth of Global Tropospheric Methane," *Geophysical Research Letters*, vol. 33, L22808, 2006, DOI: 10.1029/2006GL027330.

Chapter 2: Whodunnit?

1. F. Foukal, C. Fröhlich, H. Spruit, and T. M. L. Wigley, "Variations in solar luminosity and their effect on the Earth's climate," *Nature*, vol. 433, pp. 161–6, 2006.

2. Mike Lockwood and Claus Fröhlich, "Recent oppositely directed trends in solar climate forcings and the global mean surface air temperature," *Proceedings of the Royal Society A;* DOI: 10.1098/ rspa.2007.1880 (2007), available on the Web at http://www.pubs. royalsoc.ac.uk/media/proceedings_a/rspa20071880.pdf.

3. The effect of clouds on climate isn't quite as straightforward as this. Though low-lying clouds do reduce the temperature in this way, clouds in the upper part of the troposphere can actually warm the climate by reflecting infrared radiation back down to Earth. Also, losing clouds means that more warming gets through during the day, but less heat is retained at night. Struggling to deal with these issues,

and to predict exactly how much cloudiness—and of what sort—we can expect in the future, is one reason there is such a wide spread of model predictions for how much temperature increase we can expect, with a given increase in carbon dioxide.

4. These particles can cool the planet in two ways. In the "direct effect" they bounce sunlight directly back into space. But in the "indirect effect" they can also help to seed cloud droplets, which also reflect sunlight. If clouds form around aerosols, they tend to make smaller droplets, which live longer in the atmosphere before they grow big enough to fall as rain. Thus aerosols also tend to increase the lifetimes of clouds.

5. See, for instance, the University of East Anglia's Climate Research Unit information sheet at http://www.cru.uea.ac.uk/cru/info/volcano/.

6. IPCC WGI, chapter 9, section 9.7.

7. Some of this cooling is because of the disappearing ozone layer, which inhabits the lower stratosphere. Since ozone is also a greenhouse gas, it warms its local stratosphere, and losing it means the stratosphere cools down. However, the loss of ozone isn't enough to explain the whole trend. When the models put in both the known amount of ozone loss and the known amount of carbon dioxide and methane, the stratosphere cools by the same amount that the instruments have measured. Natural spikes occur in the temperature record from the occasional volcanic eruption, but greenhouse gases are nonetheless the drivers of the overall cooling. See, for instance, V. Ramaswamy et al., "Anthropogenic and natural influences in the evolution of lower stratospheric cooling," Science, February 24, 2006, vol. 311, no. 5764, pp. 1138–41; DOI: 10.1126/science.1122587. IPCC WGI, chapter 9, also has more on this.

8. GCM originally stood for General Circulation Models, which are the atmosphere-only forerunners of modern climate models.

9. V. Ramanathan et al., "Warming trends in Asia amplified by brown cloud solar absorption," Nature, vol. 448, pp. 575–78, August 2, 2006, DOI: 10.1038/nature06019.

10. The resumption of major volcanic activity after 1956 probably helped, but it came ten years too late to explain the onset of this cooling.

11. See, for instance, Peter A. Stott et al., "External control of 20th-century temperature by natural and anthropogenic forcings," Science, vol. 290, p. 2133 (2000); DOI: 10.1126/science.290.5499.2133, and

the accompanying commentary, Francis W. Zwiers and Andrew J. Weaver, "The causes of 20th-century warming," *Science,* vol. 290, pp. 2081–83 (2000); DOI: 10.1126/science.290.5499.2081. Also Gerald A. Meehl et al., "Combinations of natural and anthropogenic forcings in twentieth-century climate," *Journal of Climate,* vol. 17, no. 19, pp. 3721–27 (2004).

12. Brian J. Soden et al., "The radiative signature of upper tropospheric moistening," *Science,* vol. 310, p. 841 (2005); DOI: 10.1126/science .1115602.

Chapter 3: Feeling the Heat

1. David W. Inouye, Billy Barr, Kenneth B. Armitage, and Brian D. Inouye, "Climate change is affecting altitudinal migrants and hibernating species," *Proceedings of the National Academy of Sciences,* vol. 97 (4), pp. 1630–33, February 15, 2000.

2. See "Climate Change 2007: Impacts and Adaptation" (written by Working Group II and subtitled "Working Group II Contribution to the Fourth Assessment Report of the IPCC Corporate Author Intergovernmental Panel on Climate Change," and hereafter referred to as IPCC WGII), chapter 1, section 1.3.5.6, IPCC reports (Cambridge, England: Cambridge University Press, 2007).

3. J. Alan Pounds et al., "Widespread amphibian extinctions from epidemic disease driven by global warming," *Nature,* vol. 439, pp. 161–67, January 12, 2006; DOI: 10.1038/nature04246. Other researchers have suggested the decline may be due to climate-related reduction in leaf litter on the ground. See Steven M. Whitfield et al., "Amphibian and reptile declines over 35 years at La Selva, Costa Rica," *PNAS,* vol. 104, pp. 8352–56 (2007).

4. See IPCC, WGII, chapter 4.

5. J. C. Stroeve et al., "Tracking the Arctic's shrinking ice cover: another extreme September minimum in 2004," *Geophysical Research Letters,* vol. 32, L04501 (2005).

6. Gabrielle Walker, "The tipping point of the iceberg," *Nature,* vol. 441, pp. 802–5, 2006.

7. J. E. Overland, M. C. Spillane, D. B. Percival, M. Wang, and H. O. Mofjeld, "Seasonal and Regional Variation of Pan-Arctic Surface Air Temperature Over the Instrumental Record," *Journal of Climate,* vol. 17, no. 17, pp. 3263–82 (2004).

8. See Walker, "The tipping point of the iceberg," and T. C. Johns et al., "Anthropogenic climate change for 1860 to 2100 simulated with the HadCM3 model under updated emissions scenarios," *Climate Dynamics*, vol. 20, pp. 583–612, 2003.

9. IPCC, WGII, chapter 4, box 4.3.

10. Open ocean plankton might benefit from the lifting of the ice lid, but the Arctic is poor enough in nutrients that this would probably not be much compensation. See V. Smetacek, V. Nicol, and S. Nicol, "Polar ocean ecosystems in a changing world," *Nature*, vol. 437, pp. 362–68, 2005, DOI: 10.1038/nature04161.

11. "Ocean acidification due to increasing atmospheric carbon dioxide," Royal Society policy document 12/05 (2005).

12. K. Caldeira and M. E. Wickett, "Anthropogenic carbon and ocean pH," *Nature*, vol. 425, p. 365, 2003.

13. http://www.tos.org/oceanography/issues/issue_archive/issue_pdfs/20_2/20.2_caldeira.pdf.

14. See IPCC WGII, Summary for Policymakers.

15. See IPCC WGII, chapter 4, for much more on the specific services that different unmanaged ecosystems provide us with.

16. For this and following, see the discussion in IPCC WGI, chapter 9, section 9.5.4.3.1.

17. Michela Biasutti and Alessandra Giannini, "Robust Sahel drying in response to late 20th century forcings," *Geophysical Research Letters*, vol. 33, L11706 (2006), DOI: 10.1029/2006GL026067, 2006.

18. Julian Borger, "Scorched," *Guardian*, Saturday, April 28, 2007.

19. In the past thirty years, total numbers of hurricanes have decreased slightly over that time period in most parts of the tropical ocean. However, there have been unusually large numbers of hurricanes in the North Atlantic in nine of the last eleven years, culminating in the record-breaking 2005 season.

20. T. P. Barnett et al., "Penetration of human-induced warming into the world's oceans," *Science*, vol. 309, p. 284, 2005, DOI: 10.1126/science.1112418.

21. K. A. Emanuel, "Increasing destructiveness of tropical cyclones over the past 30 years," *Nature*, vol. 436, pp. 686–88, 2005. See also Emanuel's excellent essay, available at http://wind.mit.edu/~emanuel/anthro2.htm, which gives much more detail about the mechanics of hurricanes, and also points out that the increased damage felt in the United States over the past thirty years has as yet more to do with the

rush to live at the coast and flimsy building standards than the impact of the stronger hurricanes. Also note that it's possible that increasing wind strengths high in the atmosphere will help to cut up baby hurricanes before they turn into dangerous adults.

22. Christoph Schär and Gerd Jendritzky, "Hot news from summer 2003," *Nature*, vol. 432, pp. 559–60, December 2, 2004.

23. For arguments in favor of the higher figure, see http://www.earth-policy.org/Updates/2006/Update56.htm.

24. P. A. Stott, D. A. Stone, and M. R. Allen, "Human contribution to the European heat wave of 2003," *Nature*, vol. 432, pp. 610–14, December 2, 2004.

Chapter 4: In the Pipeline

1. See, for instance, Peter A. Stott and J. A. Kettleborough, "Origins and estimates of uncertainty in predictions of twenty-first century temperature rise," *Nature*, vol. 416, pp. 723–26, 18 April 2002.

2. Jeffrey A. Yin, "A consistent poleward shift of the storm tracks in simulations of 21st century climate," *Geophysical Research Letters*, vol. 32, L18701, DOI: 10.1029/2005GL023684, 2005. There's also a good entry on this subject at http://www.realclimate.org/index.php/archives/2006/12/on-mid-latitude-storms/.

3. Nicholas Stern, *Stern Review on the Economics of Climate Change* (Cambridge, England: Cambridge University Press, 2007) (hereafter referred to as the *Stern Review*).

4. Stott, Stone, and Allen, "Human contribution to the European heat wave of 2003."

5. See W. R. Keatinge et al., "Heat related mortality in warm and cold regions of Europe: observational study," *British Medical Journal*, vol. 321, pp. 670–73, 2000, available online at http://www.bmj.com/cgi/reprint/321/7262/670.

6. *Stern Review*, Box 3.5.

7. In 1957–58 it was 0.23 km^2 and by 1997 it had grown to 1.65 km^2. For this and other details of the Tsho Rolpa adaptation scheme, see IPCC WGII, chapter 17, box 17.1.

8. For this and the following, see IPCC WGII, chapter 17, section 17.2.2.

9. Keatinge et al., "Heat related mortality in warm and cold regions of Europe."

Chapter 5: Climate Wild Cards

1. The other reason is that northern Europe has a maritime climate—few countries are far from the relative warmth of the sea—whereas central parts of North America are trapped in the middle of a continent.

2. Walker, "The tipping point of the iceberg," and references therein; see also Stephen Battersby, "The Great Atlantic shutdown," *New Scientist*, April 15, 2006.

3. A compensating drop in sea level in the southern hemisphere would be less severe because it would be spread over a larger area.

4. S. Rahmstorf et al., "Thermohaline circulation hysteresis: a model intercomparison," *Geophysical Research Letters*, vol. 32, L23605, 2005, DOI: 10.1029/2005GL023655.

5. Walker, "The tipping point of the iceberg."

6. Andrew Shepherd and Duncan Wingham, "Recent sea-level contributions of the Antarctic and Greenland Ice Sheets," *Science*, vol. 315, pp. 1529–32, March 16, 2007, DOI: 10.1126/science.1136776.

7. C. H. Davis, Y. Li, J. R. McConnell, M. M. Frey, and E. Hanna, "Snowfall-driven growth in East Antarctic Ice Sheet mitigates recent sea-level rise," *Science*, vol. 308, no. 5730, pp. 1898–1901, 2005, DOI: 10.1126/science.1110662.

8. E. Domack et al., "Stability of the Larsen B ice shelf on the Antarctic Peninsula during the Holocene epoch," *Nature*, vol. 436, pp. 681–85, 2005, DOI: 10.1038/nature03908.

9. T. A. Scambos, J. A. Bohlander, C. A. Shuman, and P. Skyarca, "Glacier acceleration and thinning after ice shelf collapse in the Larsen B embayment, Antarctica," *Geophysical Research Letters*, vol. 31, L18401 2004, DOI: 10.1029/2004GL020670.

10. See, for example, Andrew Shepherd et al., "Inland thinning of Pine Island glacier, West Antarctica," *Science*, vol. 291, pp. 862–64, February 2, 2001, DOI: 10.1126/science.291.5505.862.

11. O. M. Johannessen et al., "Recent ice-sheet growth in the interior of Greenland," *Science*, vol. 310, pp. 1013–16, 2005; and W. Krabill et al., "Greenland Ice Sheet: increased coastal thinning," *Geophysical Research Letters*, vol. 31, 2004.

12. E. Rignot and P. Kanagaratnam, *Science*, vol. 311, p. 986 (2006).

13. J. M. Gregory and P. Hybrechts, "Ice-sheet contributions to future sea-level change," *Philosophical Transactions of the Royal Society of London* A, 364, pp. 1709–31, 2006, DOI: 10.1098/rsta.2006.1796.

14. H. Jay Zwally et al., "Surface melt-induced acceleration of Greenland Ice-Sheet flow," *Science,* vol. 297, pp. 218–22 (2002).

15. Gabrielle Walker, "A world melting from the top down," *Nature,* vol. 446, pp. 718–21, 2007, and references therein.

16. V. Romanovsky et al., "Permafrost Temperature Records: Indicators of Climate Change," *EOS, AGU Transactions,* vol. 83, no. 50, pp. 589–94, December 10, 2002.

17. K. M. Walter et al., "Methane bubbling from Siberian thaw lakes as a positive feedback to climate warming," *Nature,* vol. 443, pp. 71–75, September 7, 2006, DOI: 10.1038/nature05040.

18. T. Johansson et al., "Decadal vegetation changes in a northern peat land, greenhouse gas fluxes and net radiative forcing," *Global Change Biology,* vol. 12, pp. 1–18 (2006), DOI: 10.1111/j.1365-2486.2006.01267.x.

19. J. E. Hanson, "Scientific reticence and sea level rise," *Environmental Research Letters,* 2:024002; 2007, DOI: 10.1088/1748-9326/2/2/024002.

20. See IPCC WGII, chapter 7.

21. See, for instance, Jim Lovelock, *The Revenge of Gaia* (London: Allen Lane, 2006); and Fred Pearce, *The Last Generation: How Will Nature Take Her Revenge for Climate Change?* (Nottingham, England: Eden Books, 2006).

Chapter 6: What Should We Aim For?

1. These data all come from IPCC WGII, chapter 19, table 19.1, and chapter 20, table 20.7. Note that there's already a danger of confusion when anyone talks about a specific temperature rise because some of these refer to rises above the temperature of preindustrial times, before global warming had begun to take hold, while others are talking about rises relative to today, when we have already had 1.4°F of warming. When anyone mentions a specific temperature rise, it's always worth checking the reference point. Throughout this book we will always refer to rises over preindustrial times, unless we say otherwise.

2. Cited by Michael Oppenheimer and Annie Petsonk in "Article 2 of the UNFCCC: Historical Origins, Recent Interpretations," *Climatic Change,* vol. 73, pp. 195–226 (2005), DOI: 10.1007/s10584-005-0434-8.

3. "Climate Protection Strategies for the 21st Century. Kyoto and Beyond," WBGU, 2003, available on the Web at http://www.wbgu.de/ wbgu_sn2003_engl.html.

4. "Meeting the Climate Challenge," International Climate Change Taskforce, available on the Web at http://www.americanprogress.org/kf/climatechallenge.pdf.

5. http://register.consilium.europa.eu/pdf/en/05/st07/st07242.en05.pdf.

6. Because these values are so uncertain, we have rounded them to the nearest 0.5°F. These numbers are a combination of the results described in "Climate Change 2007: Mitigation of Climate Change" (hereafter referred to as IPCC WGIII), chapter 3, especially table 3.9 and the discussion in section 3.5 (Cambridge, England: Cambridge University Press, 2007), and in G. R. Harris et al., "Frequency distributions of transient regional climate change from perturbed physics ensembles of general circulation model simulations," *Climate Dynamics*, vol. 27, no. 4, pp. 357–75, 2006, DOI: 10.1007/s00382-006-0142-8. The range is for temperatures at an 80 percent confidence level.

7. *Stern Review*, chapter 7.

8. IPCC WGIII, chapter 1.

9. S. Pacala and R. Socolow, "Stabilisation wedges: solving the climate problem for the next 50 years with current technologies," *Science*, vol. 205, pp. 968–72, August 13, 2004.

10. Fred Pearce describes this well in the appendix to *The Last Generation*.

11. Unlike the wedges, which start at nothing today and increase linearly over the next fifty years, the Virgin prize requires removing a flat billion tons per year for at least a decade. If we started doing this in ten years' time and continued for four decades, that would give a total of forty billion tons, which is about half a Socolow wedge.

12. Johannes Lehmann, "A handful of carbon," *Nature*, vol. 447, pp. 143–44, 2007.

13. Oliver Morton, "Is this what it takes to save the world?," *Nature*, vol. 447, pp. 132–36, 2007, DOI: 10.1038/447132a2007.

14. P. J. Crutzen, "Albedo Enhancement by Stratospheric Sulfur Injections: A Contribution to Resolve a Policy Dilemma?" *Climatic Change*, vol. 77, pp. 211–20, 2006.

Chapter 7: More from Less

1. IPCC WGIII, chapter 6. This figure doesn't include halocarbons, which add about 1.5 gigatons CO_2eq per year, though are likely to be phased out in any case.

2. http://www.eere.energy.gov/consumer/your_home/lighting_
   daylighting/index.cfm/mytopic=11980.

3. See http://www.eere.energy.gov/consumer/your_home/space_
   heating_cooling/index.cfm/mytopic=12300.

4. Ibid.

5. IPCC WGIII, chapter 6, section 6.4.9.

6. See the IEA fact sheet at http://www.iea.org/textbase/papers/2005/
   standby_fact.pdf.

7. http://standby.lbl.gov/faq.html.

8. *Stern Review*, box 24.9, page 600.

9. Classic studies carried out by Robert Socolow at Princeton University
   in 1978 showed that the energy consumption of identical houses with
   different occupants could vary by a factor of two.

10. http://kim.foresight.gov.uk/horizon_scanning_centre/Energy/
    Potential_Role_ST/Potential_Role_ST.html.

11. Ibid.

12. Ibid.

13. See IPCC WGIII, chapter 8, section 8.3.2.

14. IPCC WGIII, chapter 8.

15. *Stern Review*, chapter 25, figure 25.1. These numbers vary from study
    to study, and WGIII, chapter 9 of the IPCC report gave a slightly
    smaller but still very substantial figure of 5.8 gigatons CO_2/yr "in the
    1990s." Chapter 3 of the same report gave the figures for emissions
    from deforestation and soil decomposition after logging as 7 to 16
    percent of 2004 total greenhouse gas emissions.

16. Selective logging can be done much more sustainably with just a
    little care. Certification showing the origin of tropical hardwoods and
    whether they had been harvested in a sustainable way would help to
    put pressure on loggers to do it right.

17. Planting trees isn't always a climate panacea. There is some suggestion
    that putting new trees in middle and higher latitudes means replacing
    bright land with dark treetops, and hence encourages Earth to soak up
    more sunlight, causing further warming. See S. Gibbard, K. Caldeira,
    G. Bala, T. J. Philips, and M. Wickett, "Climate effects of global land
    cover change," *Geophysical Research Letters*, 32 (23) L23705, 2005.
    That doesn't, however, apply to the tropics.

18. This is the result of the "top-down" models that start from the varied
    CO_2 levels in different parts of the atmosphere and work backward to
    see where they are coming from. The models that calculate from the

"bottom up" using existing land areas find lower values, though they may underestimate the ability of the forests to act as carbon sinks. See IPCC WGIII, chapter 9, for more discussion of this.

19. The extent to which forests are already acting as net carbon sinks is controversial. See, for instance, the recent *Science* paper showing that tropical forests store considerably more carbon than had previously been thought: B. Stephens et al., "Weak Northern and Strong Tropical Land Carbon Uptake from Vertical Profiles of Atmospheric CO_2," *Science*, vol. 316, pp. 1732–35 (2007).

20. *Stern Review*, chapter 25, box 25.2.

21. *Stern Review*, chapter 25, p. 607.

22. See IPCC WGIII, chapter 9, table 9.3.

23. Quoted in IPCC WGIII, chapter 10, section 10.1.

Chapter 8: Planes, Trains, and Automobiles

1. Emissions were 6.3 billion tons CO_2 eq in 2004, having risen from 1970 by 120 percent.

2. IPCC WGIII, chapter 1.

3. http://kim.foresight.gov.uk/horizon_scanning_centre/Energy/ Potential_Role_ST/Potential_Role_ST.html.

4. It also said that rising prices would increase the cost of food imports to developing countries by 9 percent in 2007. The report is available on the Web at http://www.fao.org/docrep/010/ah864e/ah864e00.htm.

5. A. E. Farrell et al., "Ethanol can contribute to energy and environmental goals," *Science*, vol. 311, pp. 506–8, January 27, 2006. The numbers in the paper differ slightly from the ones quoted here, but see also the appended erratum.

6. http://www.iea.org/textbase/techno/essentials2.pdf.

7. See, for instance, the Friends of the Earth briefing at http://www.foe .co.uk/resource/briefings/palm_oil_biofuel_position.pdf.

8. "Sustainable Bioenergy: A Framework for Decision Makers," available at http://esa.un.org/un-energy/pdf/susdev.Biofuels.FAO.pdf. See also Fred Pearce, "Fuels gold: big risks of the biofuel revolution," *New Scientist*, September 25, 2006, for an excellent overview of the ins and outs of biofuels.

9. Partly because of the potential problems in doing the greenhouse accounting for biofuels, the UN report also concluded that bioenergy could best serve the greenhouse cause if used in place of coal to

make electricity and heat. But it also acknowledged that the dearth of alternative fuels for transportation made further research on biofuels very attractive.

10. *Stern Review,* p. 388, box 15.6.

11. *Stern Review,* p. 388, box 15.6, and IPCC WGIII, chapter 5.

12. See, for instance, the analysis by the United Kingdom's Royal Commission on Environmental Pollution: "The Environmental Effects of Civil Aircraft in Flight," published in 2002 and available on the Web at: http://www.rcep.org.uk/avreport.htm.

13. IPCC WGIII, chapter 5, section 5.3.3.

14. IPCC WGIII, chapter 5, section 5.3.3, table 5.7. See also an excellent summary of the potential technological developments, along with their possible hazards, in Bennett Daviss, "Green sky thinking," *New Scientist,* February 22, 2007.

15. IPCC WGIII, chapter 5, section 5.4.2.1, and table 5.13.

16. There's an excellent discussion of these issues in a special report for the United Kingdom's *Observer* newspaper, "The big green dilemma," by Tom Robbins, July 1, 2007, available on the Web at http://www.guardian.co.uk/travel/2007/jul/01/escape.green.

17. IPCC WGIII, chapter 5, figure 5.7. Note that this shows the values in terms of carbon, rather than carbon dioxide. To convert, you need to multiply by forty-four and divide by twelve. See also V. Eyring et al., "Emissions from international shipping: 1. The last 50 years," *Journal of Geophysical Research—Atmospheres,* vol. 110 (D17), p. 17305, September 15, 2005, abstract available on the Web at http://www.mpg.de/english/researchResults/researchPublications/researchReports/GEO/200541_043.shtml.

18. See IPCC WGIII, chapter 5, table 5.4.

19. http://www.railteam.eu/.

20. http://www.acea.be/files/463-8.pdf.

21. *Stern Review.*

22. Fuel cells could also be used as a storage device to get around the fickle nature of renewable sources such as solar and wind.

23. *Stern Review,* p. 405, box 16.1.

Chapter 9: Power to Change

1. IPCC WGIII, chapter 1, p. 3.

2. IPCC WGIII, chapter 4, table 4.2.

3. IPCC WGIII, chapter 1, p. 13.

4. "IPCC Special Report on Carbon Capture and Storage," 2005, chapter 1. Available on the Web at http://www.ipcc.ch/activity/srccs/index.htm.

5. *Stern Review*, p. 400.

6. http://www.worldenergyoutlook.org/press_rel06.asp.

7. Unless otherwise specified, the information in the rest of this chapter comes either from IPCC WGIII, chapter 4, or the Foresight Horizon Scanning Technology report at http://kim.foresight.gov.uk/horizon_scanning_centre/Energy/Potential_Role_ST/Potential_Role_ST.html.

8. IPCC WGIII, chapter 4, section 4.3.3.1, for the details in this paragraph.

9. IPCC WGIII, chapter 4, table 4.2.

10. IPCC WGIII, chapter 4.

11. IPCC WGIII, chapter 4, table 4.2.

12. http://www.awea.org/projects/.

13. http://www.newwindenergy.com/wind-farms/jersey-atlantic-wind-farm/.

14. *Stern Review*, box 9.2.

15. "IPCC Special Report on Carbon Capture and Storage," 2005, available on the Web at http://www.ipcc.ch/activity/srccs/index.htm.

16. http://fossil.energy.gov/sequestration/partnerships/index.html.

17. http://www.futuregenalliance.org/.

18. *Stern Review*, box 24.7, p. 593.

19. See IPCC WGIII, chapter 4, section 4.3.6.

20. "IPCC Special Report on Carbon Capture and Storage," 2005, available on the Web at http://www.ipcc.ch/activity/srccs/index.htm.

21. The fusion process generates a few stray neutrons that will slam into the walls of the reactor and eventually make it radioactive. But the ash itself from the reaction is free of any radioactivity, and the overall waste burden is vastly less than for a fission plant.

22. http://www.worldenergyoutlook.org/press_rel06.asp.

Chapter 10: It's the Economy, Stupid

1. *Stern Review*, chapter 6. The baseline level of costs was 5 percent of GDP. Including "nonmarket" impacts on the environment and human health increased the figure to 11 percent, including the possibility that the climate may be more sensitive than we think

took it up to 14 percent, and considering that the poorer parts of the world would receive a disproportionate share gave a final figure of 20 percent. Note that these figures mark the eventual impact of business as usual, and as such wouldn't be achieved for several centuries.

2. On top of that, Stern used an unusually low figure for the discounting level between rich and poor. Put simply, this number accounts for the fact that a percentage drop in income for someone struggling to survive on £1 per day is much worse than the same percentage drop for someone earning £100 per day. Stern used a figure for this parameter of one, which means that, in this example, a benefit going to the person earning £1 a day would be valued one hundred times more than the same percentage benefit going to the person on £100 per day. A figure of two, which is more common, would change the value factor in this example to ten thousand. Other economists were indignant that Stern was not more egalitarian in his choice, saying that this, too, builds in an excessive weight for future generations. Since economies tend to grow, people in the future are likely to be richer than we are today. Giving less weight to the poor (i.e., us) and more weight to the rich (i.e., our grandchildren) also makes damage to the future seem more expensive. See William Nordhaus, "Critical assumptions in the Stern Review on Climate Change," *Science,* vol. 317, pp. 201–2, July 13, 2007. There is also a useful editorial in the *Economist,* "Shots across the Stern," December 13, 2006.

3. Nicholas Stern and Chris Taylor, "Climate Change: Risk, Ethics and the Stern Review," *Science,* vol, 317, pp. 203–4, July 13, 2007.

4. Dasgupta wrote, "To accept it would be to claim that the current generation in the model economy ought literally to starve itself so that future generations are able to enjoy ever-increasing consumption levels. (In fact, to suppose that eta is 1 is also to suppose that starvation isn't all that painful!)" See "Comments on the Stern Review's Economics of Climate Change by Sir Partha Dasgupta," available on the Web at http://www.econ.cam.ac.uk/faculty/dasgupta/STERN.pdf.

5. IPCC WGII, chapter 20, p. 16.

6. Partha Dasgupta, "Discounting Climate Change," *Review of Environmental Economics and Policy,* in the press.

7. The IPCC report roughly agrees with these figures, though the range is shifted slightly higher. It suggests that the cost for 550 ppm by 2050 would be between 1 and 5 percent of GDP; that for 650 ppm it would

be "less than 2% of GDP"; and that for 450 ppm there are too few studies to give a reliable estimate. See IPCC WG III, Summary for Policymakers.

8. Available on the Web at: http://www.mckinseyquarterly.com/article_abstract_visitor.aspx?ar=1911&L2=3&L3=0&srid=246 (requires free registration).

9. *Stern Review*, p. xviii.

10. *Stern Review*, p. 386.

11. The Kyoto scheme began operation in February of that year, when the Protocol came into force.

12. The United Kingdom, Ireland, Spain, Italy, Austria, and Greece all had emissions that were higher than their caps. But emissions from the rest of the European Union were lower. See, for instance, A. D. Ellerman and D. K. Buchner, "The European Union Emissions Trading Scheme: Origins, Allocations and Early Results," *Review of Environmental Economics and Policy*, vol. 1, no. 1, pp. 66–87, winter 2007.

13. Martin Brough, "Emission trading—can the carbon market drive investment decisions?" *Power Engineering International*, available on the Web at http://pepei.pennnet.com/display_article/297274/17/ARTCL/none/none/1/Emission-Trading---Can-the-carbon-market-drive-investment-decisions?

14. For these reasons, economists have already given the experiment a cautious thumbs-up. See, for example, the set of papers in *Review of Environmental Economics and Policy*, vol. 1, no. 1, winter 2007.

15. "State of the Carbon Market 2007," available via the World Bank's carbon Web site at http://carbonfinance.org/.

16. See *Stern Review*, chapter 17, for more on this.

17. http://assets.panda.org/downloads/emission_impossible__final_.pdf.

18. M. Wara, "Is the global carbon market working?" *Nature*, vol. 445, pp. 595–96, February 18, 2007.

19. Nick Davies, "Abuse and incompetence in fight against global warming," *Guardian*, June 2, 2007.

20. www.cdmgoldstandard.org.

21. *Stern Review*, p. 610.

22. See http://unfccc.int/methods_and_science/lulucf/items/3757.php. For more discussion of this issue, see Raymond E. Gullison et al., "Tropical forests and climate policy," *Science*, vol. 316, pp. 985–86, May 18, 2007, published online May 10, 2007 (DOI: 10.1126/

science.1136163); and M. Santilli et al., "Tropical deforestation and the Kyoto Protocol," *Climatic Change*, vol. 71, no. 3, pp. 267–76(10), August 2005.

23. *Stern Review*, p. 616, box 25.5.
24. *Stern Review*, p. 626, box 26.1.
25. *Stern Review*, p. 628, box 26.2.
26. There's an excellent summary of the business state of play in Kurt Kleiner, "The corporate race to cut carbon," *Nature Reports Climate Change*, DOI: 10.1038/climate.2007.31, available on the Web at: http://www.nature.com/climate/2007/0708/full/climate.2007.31.html.
27. http://www.pewclimate.org/docUploads/PEW_CorpStrategies.pdf.

Chapter 11: The Road from Kyoto

1. The Nuclear Non-Proliferation Treaty is one of the few that come close.
2. http://unfccc.int/files/kyoto_protocol/background/status_of_ratification/application/pdf/kp_ratification.pdf.
3. M. R. Rapauch et al., "Global and regional drivers of accelerating CO_2 emissions," *PNAS*, vol. 104, no. 24, pp. 10288–93, June 12, 2007.
4. See Ecofys report, "Factors underpinning future action," p. 13, available at http://www.defra.gov.uk/science/project_data/DocumentLibrary/GA01093/GA01093_4191_FRP.pdf (hereafter referred to as "Ecofys report"), for much more detail on this and how it depends on the means of distributing national targets. For the record, 550 ppm equivalent would mean a reduction for industrialized nations of around 70 percent by 2050, and 650 ppm equivalent a reduction of 60 percent.
5. Evidence from Friends of the Earth to the UK's Environmental Audit Committee, July 2007. See the full report at http://www.publications.parliament.uk/pa/cm200607/cmselect/cmenvaud/460/460.pdf.
6. Ibid.
7. Ecofys report fact sheets.
8. Ecofys report, p. 13. Note that the exact figures depend on population scenarios, and would probably have to be adapted to reflect actual population figures as time passed.
9. Ecofys report.
10. *Stern Review*, p. 527.

Chapter 12: Rapidly Developing Nations

1. The Ecofys report contains much more information about the data sources and their reliability.
2. See http://www.mnp.nl/en/dossiers/Climatechange/moreinfo/ Chinanowno1inCO2emissionsUSAinsecondposition.html.
3. See the World Resource Institute's report at http://www.wri.org/ biodiv/topic_content.cfm?cid=4218.

Chapter 13: Industrialized Nations

1. Cited in Oppenheimer and Petsonk, "Article 2 of the UNFCCC: Historical Origins, Recent Interpretations."
2. http://www.nrdc.org/media/docs/020403.pdf.
3. All resigned their posts in order to join the government, and the oil tanker that Chevron named after Rice in tribute to her contribution to the company was subsequently rechristened after an outcry.
4. Clinton later said, of the Bush administration: "I signed Kyoto and they got out of it." http://www.washingtonpost.com/wp-dyn/content/ article/2005/07/13/AR2005071302350.html.
5. See, for instance, the report "Atmosphere of Pressure" by the Union of Concerned Scientists and Government Accountability Project, February 2007, which documents many of these instances. It's available at http://www.ucsusa.org/assets/documents/scientific_integrity/ Atmosphere-of-Pressure.pdf.
6. See http://oversight.house.gov/story.asp?ID=1214.
7. http://cf.iats.missouri.edu/news/NewsBureauSingleNews.cfm? newsid=9842.
8. http://www.climateinstitute.org.au/cia1/downloads/AP6_Report.pdf.
9. http://dir.salon.com/story/opinion/feature/2005/08/06/muckraker/ index.html.
10. Also known as Assembly Bill 32, or AB32.
11. See http://www.rggi.org/about.htm.
12. http://www.usmayors.org/climateprotection/.
13. http://www.post-gazette.com/pg/07011/753072-28.stm.
14. http://www.rollingstone.com/politics/story/15051506/global_ warming_a_real_solution.
15. See this account from the American Council for an Energy Efficient Economy: http://www.aceee.org/press/0302carbongap.htm.

16. A fascinating insight into this process can be had by reading the heavily red-lined and highlighted document, later leaked by Greenpeace, that bears the U.S. comments on the draft proposal in Germany. See http://weblog.greenpeace.org/makingwaves/G8%20Summit%20 Declaration%20%20US%20comments%20May%2014-1.pdf.

17. http://news.bbc.co.uk/1/hi/sci/tech/6651295.stm.

18. Greg Walters, "President Has Not Decided on Kyoto," *Moscow Times,* September 20, 2003, available to subscribers at http://www .themoscowtimes.com/stories/2003/09/30/001.html. Putin did go on in the same speech to acknowledge that certain parts of the country could be hit by floods and droughts.

19. IPCC WGII, chapter 12.

20. See, for instance, Anthony Faiola, "Japanese Putting All Their Energy Into Saving Fuel," *Washington Post,* Thursday, February 16, 2006.

21. http://www.emissionstrading.nsw.gov.au/.

22. http://www.pm.gov.au/media/Speech/2006/speech2218.cfm.

23. See www.pmc.gov.au/publications/emissions/index.cfm.

24. M. R. Rapauch et al., "Global and regional drivers of accelerating CO_2 emissions." The United States has produced about 29 percent of the historical emissions, and the European Union about 26 percent.

25. To divide up the reductions among its members, the EU used the infant Triptych method (see chapter 11), which gave very wide results to reflect the diversity of countries. Luxembourg, for instance, agreed to a hefty cut of 28 percent whereas Greece was allowed to grow by 25 percent and Portugal 27 percent. An agreement modeled on this approach might now form part of the basis for a new fully international agreement.

26. http://ec.europa.eu/environment/climat/adaptation/index_en.htm.

27. http://www.consilium.europa.eu/ueDocs/cms_Data/docs/pressData/ en/ec/93135.pdf.

28. The idea is to do this via some mechanism akin to the Kyoto Protocol's Clean Development Mechanism. The same document also stated that CDM projects would continue to be linked to the European Trading Scheme even after their creator, the Kyoto Protocol, became defunct in 2012.

29. He made this statement at the global summit on sustainable development in Johannesburg in August 2002.

30. The white paper, "Meeting the energy challenge," is available for downloading at http://www.dtistats.net/ewp/.

Chapter 14: How You Can Change the World

1. http://www.foe.co.uk/resource/briefings/carbon_offsetting.pdf.
2. S. G. Gibbard et al., "Climate effects of global land cover change."
3. George Monbiot wrote an extremely articulate explanation for why planting trees is a bad way to compensate for carbon emissions, and why offsets themselves can make us dangerously complacent. His article, "Buying Complacency," was published in the *Guardian* on January 17, 2006, and is available on the Web at http://www.monbiot .com/archives/2006/01/17/buying-complacency/.
4. See, for instance, "The inconvenient truth about the carbon offset industry" by Nick Davies, *Guardian,* Saturday, June 16, 2007, http:// www.guardian.co.uk/environment/2007/jun/16/climatechange .climatechange.
5. http://www.cdmgoldstandard.org/.
6. "The validity of food miles as an indicator of sustainable development," DEFRA, 2005, available on the Web at http://statistics .defra.gov.uk/esg/reports/foodmiles/final.pdf.
7. Ibid.
8. Caroline Saunders, Andrew Barber, and Greg Taylor, "Food Miles— Comparative Energy/Emissions Performance of New Zealand's Agriculture Industry," *AERU Research Report* no. 285, July 2006, available on the Web at http://www.lincoln.ac.nz/story_images/2328_ rr285_s9760.pdf.
9. James MacGregor and Bill Vorley, "Fair miles? The concept of 'food miles' through a sustainable development lens," International Institute for Environment and Development, 2006, available on the Web at http://www.iied.org/pubs/pdf/full/11064IIED.pdf.
10. http://www.defra.gov.uk/environment/climatechange/uk/individual/ pca/pdf/pca-scopingstudy.pdf.
11. Available at http://www.ci.austin.tx.us/council/downloads/mw_ acpp_points.pdf.
12. See http://www.carbonneutralnewcastle.com/home/.
13. See http://www.capegateway.gov.za/Text/2007/7/ses_july07.pdf.
14. Manfred Milinski et al., "Stabilizing the Earth's climate is not a losing game: supporting evidence from public goods experiments," *Proceedings of the National Academy of Sciences,* vol. 103, pp. 3994–98, March 14, 2006.
15. In 2007 the *New Statesman* published an excellent article discussing the ramifications of this. See "Climate change: why we don't believe

it" by Lois Rogers, April 23, 2007, available on the Web at http://www
.newstatesman.com/200704230025.

16. For more about this, and the new technologies that might help us
achieve it, see Oliver Morton's excellent *Eating the Sun* (London:
Fourth Estate, 2007).

INDEX

THE DAVID SUZUKI FOUNDATION

The David Suzuki Foundation works through science and education to protect the diversity of nature and our quality of life, now and for the future.

With a goal of achieving sustainability within a generation, the Foundation collaborates with scientists, business and industry, academia, government and non-governmental organizations. We seek the best research to provide innovative solutions that will help build a clean, competitive economy that does not threaten the natural services that support all life.

The Foundation is a federally registered independent charity, which is supported with the help of over 50,000 individual donors across Canada and around the world.

We invite you to become a member. For more information on how you can support our work, please contact us:

The David Suzuki Foundation
219–2211 West 4th Avenue
Vancouver, BC
Canada V6K 4S2
www.davidsuzuki.org
contact@davidsuzuki.org
Tel: 604-732-4228
Fax: 604-732-0752

Checks can be made payable to The David Suzuki Foundation. All donations are tax-deductible.

Canadian charitable registration: (bn) 12775 6716 RR0001
U.S. charitable registration: #94-3204049